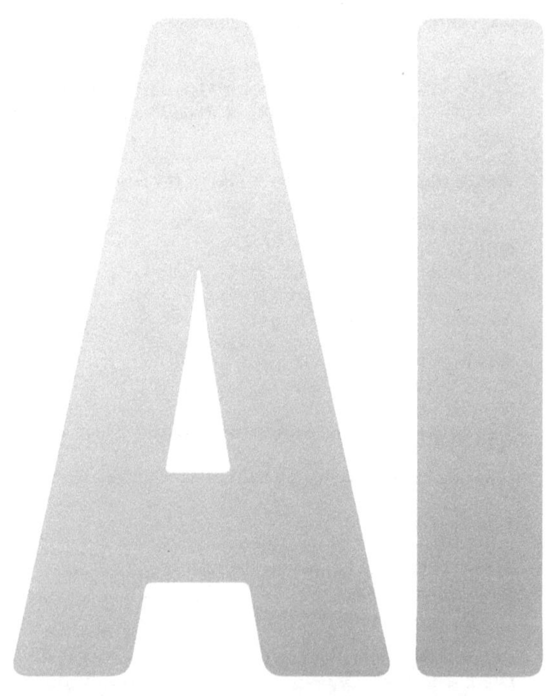

AI FOR BUSINESS

A Guide to AI Adoption

Jon Whittle

PUBLISHING

A catalogue record for this book is available from the National Library of Australia

ISBN: 9781486321421 (pbk)
ISBN: 9781486321438 (epdf)
ISBN: 9781486321445 (epub)

How to cite:
Whittle J (2026) *AI for Business: A Guide to AI Adoption*. CSIRO Publishing, Melbourne.

Published by:
CSIRO Publishing
36 Gardiner Road, Clayton VIC 3168
Private Bag 10, Clayton South VIC 3169
Australia

Telephone: [+613] 9545 8555
Email: csiropublishing@csiro.au
Website: www.publishing.csiro.au
Sign up to our email alerts: publishing.csiro.au/earlyalert

Cover: background graphic (partial) by LALAKA/Shutterstock.com

Edited by Janet Walker
Cover design by Cath Pirret Design
Typeset by Envisage Information Technology
Index by Indexicana
Printed by Ingram Lightning Source

CSIRO acknowledges the Traditional Owners of the lands that we live and work on across Australia and pays its respect to Elders past and present. CSIRO recognises that Aboriginal and Torres Strait Islander peoples have made and will continue to make extraordinary contributions to all aspects of Australian life including culture, economy and science. The use of Western science in this publication should not be interpreted as diminishing the knowledge of plants, animals and environment from Indigenous ecological knowledge systems.

Jan26_RP_ILS

Contents

A note about the use of AI in writing this book

CHATGPT WAS USED IN SOME parts of this book for editing the author's text to tighten language and improve flow. ChatGPT was used to assist the author in generating some of the tables in the book by suggesting initial content that was then edited by the author. In all cases, the author thoroughly checked any generated output for possible errors, inaccuracies, harmful content and bias. For the most part, ChatGPT was not used to generate initial versions of text. The only exception is the case study chapters in Part V, which were initially generated using the rest of the manuscript as input and were then edited by the author. Some images in this book were generated by ChatGPT for visual interest, using prompts. They are not intended to represent a particular artistic style and have been noted in the figure captions.

About the author

JON WHITTLE EARNED A PHD in Artificial Intelligence from the University of Edinburgh in 1999, where his research focused on teaching machines to solve complex mathematical puzzles. Following his doctorate, he joined the NASA Ames Research Center in California, leading AI-driven projects in space technologies and civil aviation.

Jon later transitioned into academia, teaching software engineering and AI across the United States, United Kingdom, India, Germany and United Arab Emirates. He built a distinguished leadership career, serving as Head of the School of Computing and Communications at Lancaster University and later as Dean of the Faculty of Information Technology at Monash University, Australia. From 2020 to 2025, he was Director of CSIRO's Data61, the data and digital specialist arm of Australia's national science agency.

A trusted advisor to government and industry, Jon has played a key role in shaping Australia's AI landscape. He was instrumental in developing the country's National AI Centre, which has supported over 2,000 companies in adopting AI, and has advised both federal and state governments on AI policy. In the corporate sphere, he has secured AU$30 million in strategic partnerships with global businesses and frequently consults with leading organisations on AI strategy. He sits on the Australian Institute of Company Directors' Governance of Innovation and Technology Panel. An advocate for entrepreneurship, Jon has advised numerous startups. His contributions to leadership and innovation earned him an Executive of the Year award from *CEO Magazine*.

Recognised as a thought leader in AI and digital transformation, Jon is a sought-after keynote speaker and media commentator, reaching over 4 million people annually. His podcast, *CSIRO Presents Everyday AI*, was shortlisted for an Australian Podcast Award, and he co-authored the book *Responsible AI: Best Practices for Creating Trustworthy AI Systems*.

Acknowledgements

I'D LIKE TO THANK THE team at CSIRO Publishing, especially George Knott for handholding me through the book publishing process and for curtailing the usual publishing deadlines when I was worried my book would be out of date before it hit the shelves. I'd also like to thank my editor, Janet Walker, for her patient and meticulous job.

Alex Persley first introduced me to CSIRO Publishing and was a great sounding board for many of my kookier book ideas. Thanks to Jessie Staples for providing comms support for the book, as well as for getting my popular culture references.

The team at CSIRO's Data61 inspired a lot of the thinking in this book, but I'd especially like to shout out Qinghua Lu, who quietly and diligently got on with leading a lot of the work in our responsible AI team, much of which I have referenced.

Thanks to my readers, Stefan Hajkowicz and Stefan Harrer, for reviewing the manuscript and not giving me too many corrections when I couldn't face working on the book any longer. Both have been an endless source of inspiration when discussing all things AI.

A special mention to Vered, who has always encouraged me to get my thoughts out into the world and be authentic in my voice, even when I doubted if people wanted to hear any more from me.

1

Introduction

AS A BUSINESS OR ORGANISATIONAL leader, you've probably been thinking a lot about artificial intelligence (AI) recently.

You might already be applying AI in your business, happily sitting back and reaping the rewards. Or you might be stuck, wondering exactly *how* AI can bring value to your organisation. Maybe you've tried an AI pilot but you're not sure how to scale it up. Or perhaps you're in the exploration phase, still trying to decide the best business cases for AI use. Many of you are still drowning in the deluge of AI information, wondering if it's a saviour to your business or if the whole thing's just hype.

Welcome to the no-hype zone (see Fig 1.1).

AI is everywhere. And so is the noise around it. There are hundreds of reports that tout the economic benefits of AI and just as many warning about its dangers. There is plenty of hyperbole on both sides. Depending on who you listen to, AI is either society's messiah or it will doom us. The truth, of course, lies somewhere in the middle.

AI promises a lot. McKinsey estimates that AI could add up to US$4.4 trillion annually to the global economy.[1] Similarly, IDC (International Data Corporation) projects US$20 trillion of economic value by 2030.[2] With benefits like cost reductions, productivity boosts and better decision-making it's no surprise that AI adoption is skyrocketing across industries, from healthcare to finance, manufacturing to government.

But here's the challenge: implementing AI *effectively* is far from trivial. Historically, 70 per cent of digital transformations

Fig. 1.1. No-hype zone (image generated by ChatGPT).

fail,[3] and recent data show that AI transformations are no different, with failure rates hovering around similar levels.[4] Some companies have made AI work for them, reporting organisation-wide benefits. But surveys show that most companies are still in an experimentation phase, running pilots or proofs of concept, but not yet fully integrating AI into all aspects of the business.[5]

This book cuts through the hype on both sides of the argument. It gives practical advice for business and organisational leaders, based on evidence of what works in practice. It is designed to equip you, as a leader, with the insights and strategies to get the best out of AI in your organisation and to help make sure you are one of the 30 per cent that succeed. In short, its goal is to help you become a successful *AI-savvy leader.*

THE QUESTION EVERY LEADER ASKS

I have worked with leaders across industries, giving me a front-row seat to how companies are approaching AI. One thing is clear: AI has captured the attention of executives like never before. But many are still struggling to take meaningful action. They see the potential, understand the risks and recognise that doing nothing isn't an option.

And yet, one doubt keeps them from moving forward – one question I hear more than any other.

> 'How does my organisation get started with AI?'

That question is the foundation of this book. And it's a question that the AI-savvy leader knows how to answer.

BECOMING AN AI-SAVVY LEADER

Most readers of this book *aren't* AI experts. You're a leader – of a team, a business function, or an entire organisation. Your primary role isn't technology, but AI is becoming too important for you to ignore. Just like cybersecurity, AI is no longer just the IT department's problem; it is a strategic issue that affects *every* leader. You don't need to know how to code a neural network, but you *do* need to understand AI's fundamentals and how to apply it effectively in your business.

So, what does it take to be an AI-savvy leader?

You don't need a PhD in AI but you do need to master four key areas.

1. *Know enough about AI to cut through the hype.* AI hype is everywhere. Vendors will try to sell you AI-powered solutions to any problem, even problems you don't have. To separate real value from marketing fluff, you need a solid grasp of AI's

fundamentals. Too many companies get burned because leaders expect AI to work like magic and are disappointed when it doesn't.

2. *Be clear on the reasons you think you need or want to use it, your 'why'.* AI isn't a box to check. I've seen companies set AI goals just because their board told them. Or because of a fear of missing out. That isn't a strategy. AI's value can come from increased productivity, better customer experiences or a competitive edge but either way, you need to define your purpose upfront. If you don't, you'll end up chasing AI without a clear direction.
3. *Master how you're going to go about it, the 'how'.* AI isn't just about choosing the right technology; it's about people, culture, and leadership. Change management is hard at the best of times. Digital transformation is harder. And AI transformation is harder still, because it's a high-risk, high-return technology with a lot of uncertainty. Leaders need to understand how to align their organisation, bring their teams along for the ride and create an AI strategy that actually works.
4. *See the bigger picture.* AI doesn't operate in isolation. It has ethical, legal and reputational implications. Regulatory pressures are increasing and AI's environmental and social impact is under scrutiny. AI-savvy leaders don't just think about what AI can do; they also consider what AI *shouldn't* do. Governance is key. I have always considered governance to be a very boring concept. For me, it conjures up images of iron shackles tied around an innovator's ankles. Boring or not, though, you need AI governance. AI is cool and can do cool things for your organisation. But it needs a proper governance framework.

HOW TO READ THIS BOOK

This book is in four parts, corresponding to the four ingredients to become an AI-savvy leader (Fig. 1.2).

Part I covers background on AI, including a brief history and enough, but not too much, detail on how it works. Part II discusses why you might want to use AI in your organisation, whether it's to increase productivity, add customer value, reduce cost or even disrupt your industry. Part III covers the how, with tips on developing and executing an AI strategy. Finally, Part IV discusses the bigger picture, ensuring you are informed about the ethics of AI, its impact on the environment and society and how you can set up governance structures to manage risk. An additional section brings everything together in a case study.

Each chapter tackles a critical question that leaders have about AI. I take a scientific approach, presenting the best available evidence, alongside practical advice you can act on immediately. Chapters include 'top tips' or additional information you may find

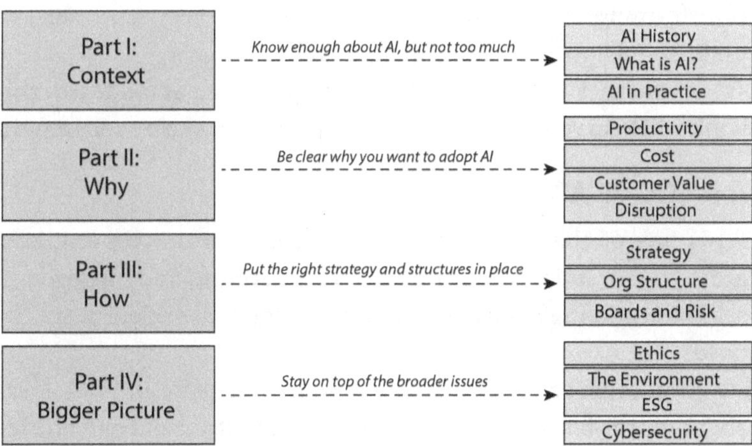

Fig. 1.2. Structure of this book.

useful. Each chapter also ends with a set of questions to prompt your thinking when adopting AI in your organisation.

You can read this book chronologically, but each chapter is designed to stand alone, so feel free to jump to the sections most relevant to you.

WHO THIS BOOK IS FOR

This book is for leaders. Whether you're a Chief Executive Officer (CEO), Chief Operations Officer (COO), board director or team manager, you'll find value here, especially if AI isn't your primary expertise but is becoming a key part of your role. Remember though that you don't need a title to be a leader. Leaders can emerge at all levels in an organisation. The book is just as suitable for experienced leaders managing hundreds of people as it is for new leaders, seeking to influence how things are done.

The book is designed for leaders across all sectors, from multinational corporations to startups, government agencies to not-for-profits. AI isn't just for tech companies. It is already transforming every industry, and leaders at every level need to understand how to leverage it effectively.

So whether you lead a global enterprise or a small team, this book will help you navigate AI with confidence, clarity and impact.

A PIECE OF ADVICE

AI is one of the most fabulous inventions of our time. Like a great friend, it will stay with you, surprise you, and push you to see the world in new ways. My advice? Enjoy the ride. I certainly have.

I first came across the term '*artificial intelligence*' in 1994. I was about to finish my mathematics degree at Oxford and had no idea what to do next. This was before the internet was everywhere, so I did what students did back then. I wandered into the university careers library and started flipping through books and magazines, searching for inspiration.

Three hours in, still lost, I was about to call it quits and head to the college bar. Then, by pure chance, I picked up a tattered magazine and flicked through its pages. Somewhere near the back, something caught my eye. It was an ad for a Master's degree in artificial intelligence at the University of Edinburgh. I'd never heard of AI before but the name alone sounded like science fiction. It was so fantastical and so full of possibilities that I wanted to learn more.

That moment changed my life. Edinburgh turned out to be one of the best places in the world to study AI. I earned my Master's, then a PhD, working on how to get computers to solve complex math equations, something which was unheard of at the time. It was exhilarating. I'd spent my childhood obsessed with numbers, and now I was teaching a machine to think like a mathematician. It felt like I was inventing the future.

One thing led to another, and soon I was at NASA, working on early AI systems for space exploration. I shared a corridor with scientists building AI-powered Mars rovers and autonomous deep-space probes. It was surreal. From there, I spent the next two decades launching AI innovation centres in the UK and Australia, eventually leading CSIRO's Data61, Australia's national research powerhouse for AI and data science.

Thirty years on from that first encounter with AI, it still fascinates me. Today, my focus is on helping organisations harness AI. Technology has advanced in ways I couldn't have imagined back in that Edinburgh classroom, but the fundamentals of AI adoption remain the same.

So as you dive in, remember this: AI isn't just a technology, it's a journey. Enjoy the ride, stay curious, and, most of all, leave the hype at the door.

Top tip: Staying up to date

When I first had the idea of writing a book on artificial intelligence, people thought I was crazy. The world of AI is changing quickly, and book publishing isn't exactly known for its break-neck speed. 'It'll be out of date before it's published,' they said.

While the technical details of AI are moving fast, the fundamentals of how to apply AI in business – the importance of purpose, people and culture – haven't changed that much since I first came across the term in 1994.

That said, the AI-savvy leader keeps on top of the latest developments. There are plenty of sources for staying up to date. I recommend Stanford University's annual AI Index,[6] which presents a wealth of data about how businesses are adopting AI and the impact on the economy. It's my go-to source for all things AI.

Part I
Context

This first part of the book covers all the background you need to be an AI-savvy leader. It starts with a brief history of AI, which goes back further than you probably think. The second chapter describes, in simple terms, what AI actually is, and arms you with the basics you need to cut through the hype. Finally, the third chapter summarises what we know about how organisations are currently using AI: in which sectors and which business functions. This chapter gives you a snapshot of where everyone else is using AI, providing some inspiration of how you should be using it.

2

Is AI new?

I HATE TO ADMIT IT, but I'm a bit of a Trekkie. Yes, I know. A computer scientist who's also a *Star Trek* fan. Not exactly original, is it?

I was seven years old when my dad took me to the cinema for the first time, to see *Star Trek: The Motion Picture* (1979). I was mesmerised. The idea of space exploration fascinated me (I would later go on to work for NASA), but what really intrigued me was something else: the artificial intelligence of the ship's computer.

The most famous *Star Trek* starship, the 'Enterprise', was shown to be intelligent in the earliest episodes of the franchise. Crew members would talk directly to the ship's computer. And the computer could understand. There was no need to type on a keyboard and no need to write a set of instructions in a computer programming language. As a budding computer scientist, *Star Trek* sparked my desire to help create a world where humans and computers could communicate seamlessly, just like they did on TV.

The original *Star Trek* series, which aired in 1966, introduced the idea of natural human–machine communication. In one episode, Captain Kirk interrogates a notorious smuggler, using the *Enterprise*'s computer as a fact-checker. By the time I was a teenager, speaking to a computer had become a regular feature in *Star Trek: The Next Generation*. Characters routinely interacted with the ship's AI, always beginning with the simple command, 'Computer', followed by an instruction. The ship's computer could do almost anything. 'Computer. Tea. Earl Grey. Hot,' said Captain Picard in nearly every episode. Or, in more dramatic moments, 'Computer. Begin auto-destruct sequence.'

What fascinated me wasn't just that the computer could talk. It was *how* it talked. Unlike the fluent, back-and-forth conversations we see with AI today, the 'Enterprise' relied on short, precise commands. That felt far more realistic to me, given the limitations of AI at the time.

Hollywood had already imagined more advanced AI interactions, of course – most notably *HAL 9000* in Kubrick's *2001: A Space Odyssey* (1968). But that level of intelligence felt too futuristic, too far removed from reality for the time. The simple grammar-based interaction in *Star Trek*, on the other hand, seemed achievable to me. I could envision a future where computers responded to basic spoken instructions. It felt within reach. But true, human-like conversations with machines? That, I thought, wouldn't happen in my lifetime.

Of course, I was very wrong.

THE BIRTH OF AI

When ChatGPT launched in November 2022, humans could hold natural conversations with AI on a mass scale for the first time. ChatGPT became the fastest-adopted technology in history. Today, over 400 million people converse fluently with AI every week.[7] That number is still smaller than the number of people that have ever watched an episode of *Star Trek*, but it's pretty impressive nevertheless.

So was AI born in 2022 with the launch of ChatGPT? Not by a long shot. AI has been around *much* longer than that. ChatGPT's rise to prominence made AI feel like a recent breakthrough, but the field's origins stretch back decades, long before the term 'artificial intelligence' even existed.

Computer scientist Alan Turing, renowned for his code-breaking work at Bletchley Park during the Second World War, in a seminal paper written in 1950 posed a profound question: 'Can machines think?'[8] He introduced what later became known as the Turing Test, a method for assessing machine intelligence.

The Turing Test works as follows. Imagine you're having a conversation with an unseen partner. You can ask anything, and they can respond however they like. If, based solely on the exchange, you cannot determine whether you're speaking to a human or a machine – and it turns out to be a machine – then the machine has passed the test.

At the time Turing wrote his article, the idea that a computer could ever pass such a test seemed far-fetched. Yet today, we engage daily with AI systems that arguably do just this.[9] Even in AI's earliest days, however, humans were surprisingly easy to fool.

In 1966, computer scientist Joseph Weizenbaum created *ELIZA*, one of the first programs capable of mimicking human conversation. It was ground-breaking for its time. One version of *ELIZA*, designed as a rudimentary psychotherapist, asked open-ended questions to help users explore their thoughts. The illusion was so convincing that Weizenbaum's own secretary asked him to leave the room so she could speak with *ELIZA* in private.

It was in 1956 that this phenomenon was formally named. At a historic conference at Dartmouth College, a group of computer scientists led by John McCarthy coined the term 'artificial intelligence' to describe a field that was beginning to take shape. Around the same time, pioneers like Frank Rosenblatt were laying the mathematical foundations for today's AI, developing early neural networks such as the perceptron.[10]

In the decades that followed, AI's progress has been anything but linear. Breakthroughs and hype cycles have repeatedly given way to stagnation and disillusionment. The field has oscillated between periods of intense excitement – AI 'summers', and prolonged setbacks – AI 'winters.' The pattern follows the classic Gartner Hype Cycle: *A new AI breakthrough emerges! This will change everything! Wait ... it's not as powerful as we thought. AI is overrated. Let's move on.*

Yet despite these cycles, AI has steadily advanced. And today, with the rise of large-scale machine-learning models, the AI summer sun is burning bright!

THE SEASONS OF AI

Since the 1950s, AI has gone through two major winters (Fig. 2.1). The first hit in the 1970s. The preceding decade had seen significant investment in AI, leading to breakthroughs in early machine translation from Russian to English, computers that could play checkers or draughts, the first industrial robot, and even machines capable of performing well on IQ and algebra tests. Optimism was high.

However, various reports in the 1960s began to sow the seeds of doubt. In 1966, a negative report on the prospects for machine translation was released,[11] and, three years later, Marvin Minsky and Seymour Papert published a book highlighting major

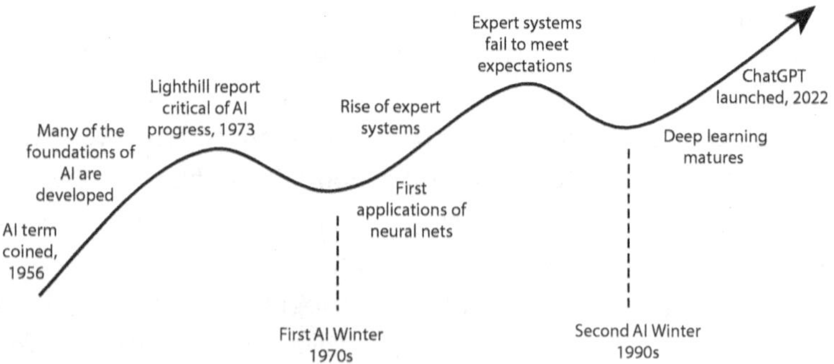

Fig. 2.1. The ups and downs of AI.

limitations of Rosenblatt's perceptron (the pre-cursor of today's neural networks).[12] The death knell came in 1973 when the British Science Research Council published the Lighthill Report,[13] a damning and pessimistic account of AI research, which led to cuts in UK funding for AI. The first AI winter was upon us.

In the 1980s, AI made a comeback. It was a heady time for AI that saw the first practical applications of neural networks and the first AI systems that could write computer programs by themselves. The big game in town, however, was something called *expert systems*. In an expert system, engineers encode a set of rules into a machine and the machine uses a combination of these rules to solve a particular problem. Expert systems showed great promise in fields such as medical diagnosis. By encoding medical knowledge as a set of rules, the machine could diagnose a patient's condition based on their symptoms.

It was a bold vision, but AI once again fell short of expectations. Expert systems struggled with real-world complexity, and concerns grew over the ethics of handing medical decisions over to machines. By the early 1990s, funding dried up and the second AI winter had begun.

It is important to note that, despite these setbacks, AI continued to make progress throughout the AI winters. In fact, some of AI's most meaningful advances have occurred in these quieter periods. With less pressure to produce flashy breakthroughs, engineers could focus on making AI work in practice. AI winters have seen AI technologies integrated into many life-changing applications.

Since the last AI winter in the 1990s, progress has been nothing short of extraordinary. AI now underpins many of the systems we rely on daily, from web search and navigation to social media and voice assistants. Today, half the US population interacts with AI at least once a day.[14] We have seen AI defeat world chess champions, triumph on live television quiz shows, power robotic explorers on

Mars, and tackle some of science's hardest problems. And all this happened before ChatGPT even existed.

AI GROWS UP

AI really grew up in the 2010s.

This decade saw the widespread adoption of *deep learning*, a technological advance that enabled neural networks to scale up dramatically, learning from vast amounts of data. Scaling up AI had been a fundamental challenge since neural networks were first conceived in the 1940s.

Two other critical developments fuelled AI's rapid progress during this time. First, cloud computing became a mature technology, allowing huge datasets to be stored and processed efficiently in virtual data centres. Second, AI needs data. Enormous amounts of the stuff. Cloud computing solved the storage and computational challenges but the data had to come from somewhere. By the 2010s, companies were collecting data at an unprecedented scale, and the explosion of social media generated an endless stream of user content. Today, the world generates 400 million terabytes of data every day,[15] the equivalent of 400 billion copies of the *Encyclopedia Britannica*. Astonishingly, 90 per cent of all the data in existence has been created in just the past two years.[16]

AI NOW – HEATWAVE OR IS WINTER COMING?

In the last few years, since the launch of ChatGPT, AI has been hot. Very hot. New AI products are released almost daily, all powered by a new brand of AI – generative AI – that underpins ChatGPT and similar tools. Generative AI can create books for you based on a simple instruction. It can create artwork and even video. It can automate parts of your business, and, if you give it access to your company data, it can make decisions on your behalf.

However, there's a good deal of scepticism about the true business value of generative AI, and it remains to be seen to what extent it translates into competitive advantage. There's also a good dose of concern about the ethical implications of generative AI for society. Gartner now ranks generative AI as having passed the point of peak hype and heading towards the trough of disillusionment.[17]

Whether we are heading for a third AI winter remains to be seen, but either way, there's no doubt that AI will continue to advance, and its impacts on business and society in the next few decades will be unprecedented.

QUESTIONS

- What AI tools do you use in daily life? Ask your colleagues and compare how their use of AI compares to your own.
- What worries or excites you the most about the evolution of AI in the coming years?

3

What is AI?

I'VE NEVER BEEN A FAN of the term 'artificial intelligence'. AI is commonly defined as a 'computer system capable of performing tasks that normally require human intelligence'. But this definition is problematic in several ways.

First, it's an inherently circular definition. It defines AI by comparing it to human intelligence. But philosophers have been arguing about human intelligence for centuries, and we don't yet have a clear consensus on what 'human intelligence' means. After all, what exactly qualifies as *intelligence*? If there isn't a precise definition of human intelligence, does it make sense to frame artificial intelligence this way?

Second, a phrase in the definition that bothers me is 'normally require'. It suggests that the definition of intelligence itself evolves over time – which, I suppose, is true. Before the invention of calculators, multiplying large numbers was considered a sign of intelligence. Yet today, no one would claim that a calculator is an example of AI. So the definition allows for moving of the goalposts, which isn't very helpful, in my view.

Third, part of the problem when defining AI is that intelligence isn't a binary notion; it is a spectrum. You can't really say if something is intelligent or not, but you can say that one thing is *more intelligent* than another. For example, is ChatGPT intelligent? Maybe, or maybe not. It can seem intelligent, but then there are lots of easy things it can't do. The answer is open to debate. But there's no debate if I say that ChatGPT is *more intelligent* than a calculator. This is self-evident.

Fourth, intelligence depends on the task we are talking about. If the task is multiplication of large numbers, a calculator is more intelligent than a 10 year old. If the task is critiquing the ideas of twentieth-century philosophers, I dare say ChatGPT is still more intelligent than a 10 year old. But if the task is to navigate a tricky series of obstacles in a forest, most AI robots will fail, whereas a 10 year old will succeed (assuming they want to!) So, on this task, the 10 year old is more intelligent.

Finally, the other problem with defining AI in terms of human intelligence is that it anthropomorphises technology. This leads to unrealistic expectations of what AI can do and fear about what it *might* do. Modern-day AI systems aren't sentient; they're just large-scale statistical pattern-matching machines. And as University of California, Berkeley psychologist Alison Gopnik

Fig. 3.1. It's maths, not magic! (Image generated by ChatGPT.)

once told me, if we'd just called it that, no one would be running around saying, 'Oh no! The large-scale statistical pattern-matching machine is coming for our jobs!'

Given all these problems then, how should one define AI?

Before I answer this question, let me be clear on one thing that AI is *not*: it's not magic. It might *feel* like magic. It *feels* magical when you ask a question to ChatGPT and it gives you an answer. It *feels* magical when AI rewrites your homework for you in perfect English. It *feels* magical when you ask AI to conjure up an image of a magician doing maths on a blackboard (Fig. 3.1). Arthur C. Clarke once said: 'Any sufficiently advanced technology is indistinguishable from magic.' But AI is not magic. It's just maths.

TWO WAYS TO THINK ABOUT AI

The truth is that artificial intelligence is a loosely defined concept. It is a catch-all term, capturing lots of different techniques: machine learning, expert systems, neural networks, large language models, and many more.

As a business or organisational leader, it can be daunting to try to understand all the different kinds of artificial intelligence that have been invented over the last 70 years. Don't worry, you don't have to. Whilst the list of AI concepts is seemingly endless (Table 3.1), all you really need to understand is that there are essentially two kinds of artificial intelligence: rule-based and data-driven AI. All types of AI fall into one of these two categories.

Let me explain the difference between rule-based and data-driven AI by using coffee as an example.[18]

I happen to like coffee. And I drink far too much of it. I like my coffee black and strong and always drink it that way. I have no idea if you, dear reader, like coffee or what kind of coffee you drink.

Table 3.1. Common AI terms and their definitions.

AI term	Definition
Machine Learning (ML)	A subset of AI that enables systems to learn from data and improve performance without explicit programming.
Deep Learning	A subset of ML that uses neural networks with multiple layers to analyse complex data patterns.
Neural Network	A computational model inspired by the human brain, consisting of layers of interconnected nodes (neurons).
Natural Language Processing (NLP)	A branch of AI that focuses on enabling computers to understand, interpret and respond to human language.
Computer Vision	A field of AI that enables machines to interpret and make decisions based on visual data from the world.
Expert System	An AI system designed to simulate human expert decision-making in a specific domain using a knowledge base and inference engine.
Generative AI	A type of AI that can create new content, such as text, images and audio, based on learned patterns.
Reinforcement Learning (RL)	A type of ML where an agent learns to make decisions by receiving rewards or penalties for its actions.
Supervised Learning	An ML technique where a model learns from labelled training data.
Unsupervised Learning	An ML technique where a model identifies patterns and relationships in unlabelled data.
Semi-supervised Learning	An ML approach that combines both labelled and unlabelled data for training.
Explainable AI (XAI)	A set of techniques that make AI decisions transparent and interpretable by humans.
Edge AI	The deployment of AI algorithms on local devices rather than centralised cloud servers.

continued

Table 3.1. continued

AI term	Definition
Federated Learning	A distributed ML approach where multiple devices train a model collaboratively without sharing raw data.
Transfer Learning	A technique where a pre-trained model is adapted for a new but related task.
Foundation Model	A large-scale AI model trained on diverse data that can be applied to multiple tasks, such as text generation, image analysis and code generation.
Large Language Model (LLM)	A type of foundation model trained on vast amounts of text data to generate human-like text.
Agentic AI	An AI system capable of autonomous decision-making, problem-solving and taking actions to achieve specific goals without constant human intervention.

But, for illustration purposes, let's suppose that you love coffee as much as I do, but that you're a bit of a coffee chameleon, choosing a different coffee type depending on what time of day it is, who you're with and how you're feeling.

Now suppose I want to create an AI system that can predict your next coffee order depending on the situation. I'm not sure why I would want to do this but maybe I run a coffee shop and want to offer excellent customer service by having your order ready before you enter the shop. My magic AI system is going to make me the most popular coffee shop in town!

We'll call my AI coffee-predicting system BaristAI.

There are two ways that BaristAI could predict your order: rule-based AI and data-driven AI (Fig. 3.2). In both cases, before BaristAI is able to predict your next coffee, I will need to *train* it. The way I train BaristAI differs depending on which form of AI I choose to implement.

a) 1. if (early morning) and (weekday)
then black coffee

2. if (weekend) and (brunch time) and (with friends)
then cappuccino

3. if (late afternoon) and (feeling tired)
then double espresso

4. if (before noon) and (with friends)
then macchiato

5. if (Saturday) or (Sunday)
then weekend

6. if (brunch time) or (early morning)
then before noon

7. if (after 11am) and (before noon)
then brunch time

b) Mon Sep 23 at 7.32am. At Giovanni's coffee shop on the corner of Broad & Meadows. You were alone, feeling blue. —*label*→ you ordered black coffee

Tues Sep 24 at 3.30pm. At Mimi's coffee shop in Fair St Mall. You were with Tess, feeling tired. —*label*→ you ordered double espresso

Wed Sep 25 at 7.52am. At Giovanni's coffee shop on the corner of Broad & Meadows. You were with Tess, feeling pumped. —*label*→ you ordered macchiato

Sat Sep 28 at 11.02am. At Mimi's coffee shop in Fair St Mall. You were with Tess and Dom, feeling refreshed. —*label*→ you ordered cappuccino

lots more labelled data....

Fig. 3.2. Rule-based (a) and data-driven (b) AI.

RULE-BASED AI

In the rule-based approach, I train BaristAI by giving it a set of explicit rules that capture what kind of coffee you tend to drink in different situations. So, for example, if you like to drink black coffee first thing in the morning, except at weekends, a cappuccino at brunch with friends on the weekend, and a double espresso if it's late afternoon and you need a pick-me-up, then I can write a set of rules that capture these preferences (Fig. 3.2a). I could obtain these rules, for example, by sitting down with you and interviewing you. Once I have enough of these rules, I can ask BaristAI for a prediction. Given any situation, BaristAI will try to match the situation against the set of rules I've given it and, using a combination of those rules, it will tell me your coffee order. Magic!

Note that, although rule-based AI seems very simple, the set of rules can get very large and complex very quickly. In real-world systems, there may be thousands, or even hundreds of thousands, of rules. This is where the AI comes in – in the form of an algorithm that searches through all these rules. The rules may be overlapping, incomplete, or even contradictory, so BaristAI needs to make a choice which rules it will use for a prediction. For example, in Fig. 3.2a, if it's 7 o'clock on a Monday morning and you're with friends, BaristAI needs to choose between rules 1 and 4. In other words, unlike traditional programming, where only one rule applies at a time, the outcome is *non-deterministic*. If you execute a traditional computer program, you will always get the same answer given the same starting point. But a rule-based AI system will need to choose an answer from multiple possibilities – it could do this randomly, or by using a set of guiding heuristics.

Rule-based AI was the basis of medical diagnosis expert systems that became popular in the 1980s. AI designers came up with hundreds of rules that could mimic a doctor's reasoning process. For example, one rule might be: If the patient has a runny nose and is sneezing and has a sore throat, then the patient may have a common cold. The reason why AI is needed in expert systems is that diagnosis is incredibly complex. Yes, the patient may have a cold, but they might also be suffering from hay fever. In other words, medical diagnosis is non-deterministic. Doctors get around this by using intuition and experience. Expert systems approximate this by using probabilities in the rules: *as an AI program, I'm 60% sure you have a cold.*

One of the clever aspects of rule-based AI is that the AI can still predict the right answer even if there isn't a single rule that matches the current situation. In Fig. 3.2a, for example, there is no explicit rule that tells BaristAI that you like a cappuccino at 11.30 on a Saturday morning when you're hanging out with your buddies. But BaristAI can figure it out by chaining together rules 2, 5 and 7. This is a form of reasoning, and it's one of the strengths of rule-based AI.

Unfortunately, rule-based AI has significant weaknesses, mainly that it takes a lot of effort to write down all the rules for real systems. It is too hard to miss one and that could throw the AI right off. For many real-world applications, it's not even possible to write down the full set of rules because we simply don't know them all. For example, you can easily define a few rules that identify the features of your cat (whiskers, long tail, silky fur, bad attitude, etc.), but you probably can't come up with a set of rules for predicting exactly how your cat will move around your apartment.

Rule-based AI was the dominant form of AI until the 1990s. It is still used today, but we now rely more commonly on data-driven AI.

DATA-DRIVEN AI

To predict your choice of coffee, a data-driven AI doesn't need an explicit set of rules. Instead, I can train BaristAI by following you around and taking notes every time you order a coffee. For each order, I write down where you are, who you're with, what time and day it is, the month of the year, and even how you're feeling based on your facial expressions. If I feed enough data into BaristAI, it can look for patterns. BaristAI would need a lot of data for this to work. Depending how addicted you are to the black stuff, I might need to follow you around for a few weeks or even months. But eventually, BaristAI will have enough data not to cover all possible scenarios but to be able to make informed guesses about your coffee-drinking habits.

There are many different forms of data-driven AI but one of the most common is *supervised learning*, where the data are explicitly labelled with an expected outcome (Fig. 3.2b). In this case, the data I give to BaristAI include information on place, time, company and emotion and are labelled with the coffee you ordered in that situation. This initial set of labelled data is called the *training data*. The information is fed into a machine-learning system (e.g. a neural network). Once it is trained, BaristAI can predict what kind of coffee you will order, even for situations it hasn't seen before.

These kinds of data-driven AI systems are incredibly powerful. Most AI systems you have come across use data-driven AI in some form. Whilst early machine-learning systems suffered from scalability issues, the increased availability of relatively cheap, high-performance computing and improvements in the efficiency of the underlying algorithms mean that modern data-driven AI systems can process billions of data points and scale up to handling highly complex problems.

There are some drawbacks of data-driven AI systems, however. First, you need high-quality data, and usually lots of the stuff. The computer science phrase 'garbage in, garbage out' sums this up. If you don't have high-quality data, you're unlikely to get a high-quality solution. You also need to make sure your data are representative of the answers you are looking for. If you only have data about tea drinkers, you won't be able to predict coffee orders. This can lead to significant ethical issues when AI developers don't understand the limitations of their data. For example, many early AI systems were found to be racially discriminatory because the datasets didn't contain much information on ethnic minority groups.

The other drawback of data-driven AI is that the AI can't explain why it made a prediction. This is often known as the *black box problem*. Think of data-driven AI as a massive box containing billions of data points. Inside the box, there is a complex web of nodes and connectors moving billions of data points around to optimise the best configuration of nodes and connectors to predict the right answer. You can look inside this box but it's so complex that you won't be able to understand what is happening.

It is the black box nature of AI that can sometimes feel like magic. Somehow, the AI analyses billions of data points and their connections to come up with answers that we weren't expecting. It can feel miraculous. But it's maths, not magic.

The black box problem is a long-standing challenge in AI. Organisations are reluctant to use AI for decision-making if they don't know how the decision was made. You wouldn't want an AI to give you a medical diagnosis if it can't justify it. Recent years, however, have seen significant progress in tackling this challenge. Some AI systems can now explain their 'reasoning', which is AI speak for a system that can provide a human-readable account of how the AI approaches a problem. It is important to note that this form of reasoning doesn't capture all the ways that humans reason, but it is a major step forward. And while it doesn't completely solve the black box problem, reasoning does go some way towards AI systems that can explain their decisions.

GENERATIVE AI

Rule-based and data-driven are the two basic types of AI. Every AI system ever invented uses one, or both, of these.

In recent years, a new AI term has risen in popularity: '*generative AI*'. Generative AI simply means AI that generates something, whether that's text, an image, audio or video. Although the focus is on generation, generative AI is a form of data-driven AI and works in just the same way. ChatGPT, the most widely known generative AI system, analyses huge amounts of text to look for patterns that enable it to learn what sentences look like so it can generate them. The principle is the same whether the AI is writing a story, creating a picture or ordering coffee.

Let's look in more detail at how ChatGPT works. I use ChatGPT as an example but note that all AI systems that generate text work in a similar way.

ChatGPT, like other conversational AI chatbots is built on what's called a *large language model* (LLM). At its core, an LLM takes an incomplete sentence and tries to predict what word will come next. To do this, the LLM comes up with a list of possible next words, based on patterns it has seen before, and it calculates the probability that each word continues the sentence (Fig. 3.3). But where does this probability come from? It is based on the text data on which the LLM has been trained: web pages, digitised books, and other sources. If the LLM can find a pattern in its dataset that matches completion of the sentence with a word, the word gets a high probability; otherwise, the probability assigned is low. Hence, an LLM is a data-driven AI, where the data is text, and the prediction task is to add a word to a sentence. The LLM

Mid-afternoon, you always order a double

	probability of next word
espresso	0.34
shot	0.22
whiskey	0.08
scoop	0.05
portion	0.02
cheeseburger	0.005

Fig. 3.3. Next-word prediction in large language models.

does this one word at a time, so it simply repeats the process until it decides to end based on some predefined halting metric.

One of the interesting aspects of LLMs is how the LLM picks the next word based on a list of possible words and their probabilities. Intuitively, we would expect the LLM always to pick the next word with highest probability. But it turns out that this approach leads to mostly boring sentences. This makes sense, as the LLM will then produce the most common possible sentence completion, and most common usually isn't that interesting. For this reason, the LLM has a parameter, called *temperature*, that determines how risky it should be in choosing the next word – in other words, how low a probability it should go with in selecting what comes next. The riskier it is, the more 'creative' the output will be. And this temperature can usually be modified by the user to produce a safer or more creative output, depending on the context.

FINE-TUNING

There is one more aspect of LLMs that I would like to cover. And that is something called *fine-tuning*. If a LLM is simply trained on billions of text pages, it might not be very useful. When given a prompt, because of the choice of probability for each word, there are hundreds of possible ways that the LLM might complete a sentence. All of them would make sense grammatically but they would have different styles, different lengths and different meanings. In essence, all you would have is a generic LLM that is very good at putting grammatically correct sentences together, but probably wouldn't write like a human does.

Fine-tuning adds extra data to improve (or fine-tune) the results of this generic LLM. In some cases, humans are used in fine-tuning. For example, a pool of hundreds of humans might be asked to rate the outputs of the generic LLM, scoring them for how human they sound. These scores are then fed back into the LLM and the result is that the LLM sounds more human next time. Sometimes, humans are not involved in fine-tuning. For example, many companies fine-tune a generic LLM on their corporate data. The result is that the LLM is good at answering questions specifically about that corporate data rather than giving very generic answers.

Top tip: The clever bit in LLMs

In a large language model, choosing the next word given a set of probabilities is easy. It doesn't require any intelligence; it's just statistics. The clever bit in a LLM is computing those probabilities in the first place.

Amazingly, it turns out that it is possible to create a model, or approximation, of these probabilities for all possible combinations of words. This is achieved using an AI technique called a *neural network*. In this instance, a neural network can be thought of as a gigantic mathematical formula that computes the probability for any given combination of words. Like any formula, a neural network has *parameters*. The challenge in predicting the next word is choosing values for all the parameters that best approximate how words are combined in everyday use.

As an analogy, imagine a set of dots on a piece of paper. Suppose you wish to draw a straight line through the dots to best approximate them (Fig. 3.4). The formula for a straight line has two parameters: one determines the angle of the line, the other determines the vertical placement of the line. By choosing the optimal values for these parameters, we end up with a line that is a good approximation of the dots.

The formula for a neural network in a LLM typically has billions of parameters. So choosing the best values for all parameters is a major challenge. This is where training of the neural network

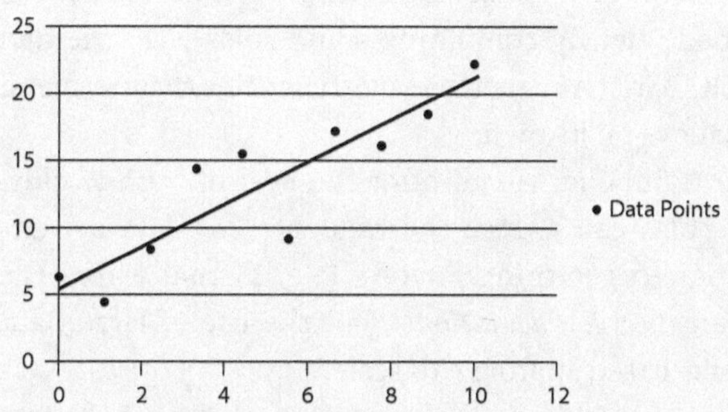

Fig. 3.4. Choosing parameter values to fit data.

comes in. In simplest terms, the neural network is trained on billions of pieces of text data. Each new data point gives the neural network a bit more information about how words combine. And each time, the network adjusts its parameters to best fit all the data it has been trained on so far. Eventually, given enough data and time, the neural network can figure out values for all the billions of parameters in a way that gives a good approximation of how language works.

It is a lot more complicated in practice, of course, but this is the essence. For readers interested in a more detailed but still accessible account of how LLMs work, I recommend Stephen Wolfram's blog.[19]

A FINAL WORD – NEUROSYMBOLIC AI

Rule-based and data-driven AI are the two types of AI. They have been presented in this chapter as alternatives but they are commonly combined. This is called *neurosymbolic AI* and some commentators believe it will eventually become the dominant form of AI. As an example, BaristAI could be implemented by combining a data-driven system that looks for patterns in your coffee habits augmented with a set of rules capturing some, but not necessarily all, of what we know about coffee. For example, there might be an explicit rule that says you never drink coffee just before bedtime. By combining some rules with the data-driven approach, BaristAI can achieve efficiencies that neither approach could achieve on its own.

Neurosymbolic AI can offer the best of both worlds. Having explicit rules can reduce the tendency for data-driven systems sometimes to get things wrong (e.g. factual errors that LLMs generate called *hallucinations*), but the use of large datasets can help a rule-based approach to scale.

I have only skimmed the surface of how AI works in this chapter. My aim was to give you enough detail, as a business leader, to understand the basic concepts. As a non-technical leader, though, you don't need to become an AI guru. After all, there have been 70 years of research and development in AI. You can't be expected to know all of that. And you don't have to.

QUESTIONS

- It is important for leaders to be able to communicate to staff clearly. Can you explain AI, in simple terms, to anyone in your organisation? Practice on a member of your family. Grandparents or young children are good for this. Use words that resonate with them.
- Can you explain all the AI terms in Table 3.1? Practice with your co-workers. Quiz each other, asking for a definition. Make it fun. I'm sure there's even a drinking game in there somewhere!

4

How is AI used in practice?

THERE IS AN OLD JOKE that AI is like teenage sex: everyone talks about it, but no one's doing it.

It's a joke that resonates because most organisations nowadays are aware of the potential benefits of AI, but relatively few have managed to transition from pilot projects to enterprise-wide adoption. This is particularly true for generative AI. The magic of generative AI means that businesses can *imagine* adopting it but haven't yet done so at the organisational scale. As a result, most companies are stuck in the experimentation phase with generative AI.

Nevertheless, the joke gives a false impression. Whilst it is arguably accurate for generative AI, more traditional forms of AI have been used in business for decades. The difference is that they are often hidden in plain sight. Much of AI happens behind the scenes – inside search engines, fraud detection systems, recommendation algorithms, logistics platforms. It's not that no one is using AI, it's that most people don't realise that they are.

Ask your grandmother if she uses AI and she's most likely to say no. But if she's ever searched the web, she's using AI. It's the same in organisations. AI is there – in credit risk models, in automated supply chains, in customer targeting systems – but it is often tucked away in specialist functions, invisible to the rest of the business.

So if AI is like anything, it's not teenage sex. It is more like an extra-marital affair. Lots of people are doing it, but no one talks about it.

FROM SPECIALIST TOOL TO UBIQUITOUS PLATFORM

For most of its history, AI has been a specialist tool, used by expert teams for narrow, high-value tasks. Take NASA, for instance, where I used to work. In the early 2000s, they launched Deep Space 1, a spacecraft designed to test out a new ion propulsion system.

For most organisations, demonstrating one new technology would be sufficient. Not for NASA, who decided to include a dozen other experimental technologies on Deep Space 1, including one very close to my heart: artificial intelligence. The benefits of AI in space are clear. It is difficult to communicate with a deep space craft from Earth. It takes hours for a signal to travel from mission control. That makes it hard to fix if something goes wrong. You can't just call a handyman. So Deep Space 1 was fitted with an AI planning and control system designed not only to detect errors in operation but also to figure out what went wrong and take corrective action. It was an experiment way before its time. And no one knew if it would work.

I was lucky enough to be in the control room when the Deep Space 1 AI was turned on. It was a tense moment. You've seen those moments in movies. Picture the scene. A bunch of technicians sit in a cavernous room, each hunched over a microphone and a computer. 'Houston, we have a problem.' And the nerds get to work, inventing ingenious solutions to some space problem that only they can solve. They don't quite know if their solution is going to work. But they try anyway. It's a matter of life and death. They set things in place. Then they flick a switch. A brief pause. It worked! The whole room explodes with cheers and back slapping.

This is what it was like for me when Deep Space 1 didn't fall out of the sky when its AI was switched on. Or at least it felt like that in my head. The only difference was that it wasn't a bunch of technicians that fixed the problem. An AI did it all by itself.

NASA is a good example of how AI has traditionally been used: powerful, but narrow applications. Robots on Mars, autonomous spacecraft, automated air traffic control. These are remarkable achievements, but they're isolated to specific corners of NASA's business. Back then, most of NASA employees – in human resources (HR), finance, marketing – weren't using AI at all.

That is now changing.

THE GENERATIVE SHIFT

What's different about generative AI is not just the technology itself, but its generality. Traditional AI requires data scientists, careful model training, and targeted deployment, all designed to solve one specific problem. Generative AI, by contrast, is far more general and can be applied to tasks across the whole of an organisation. This means AI is no longer confined to a single-use case. It is moving into the workflows of marketers, the dashboards of HR teams, the notebooks of product designers. AI is shifting from being one tool amongst many, used for specific purposes by some businesses, to being a Swiss Army knife used by all businesses for just about anything you can imagine. That's the promise, anyway.

Still, it's early days. Generative AI today is like a rocket that's cleared the launch pad but hasn't yet reached orbit. Some companies are soaring, but most are still figuring things out. According to survey data from McKinsey, only 1 per cent of company executives describe their roll-out of generative AI as mature.[20] Adoption is highest in tech, media, and finance; lowest in manufacturing, strategy, and human resources. And even where AI is being used, it's often in pockets – experimenting here, automating there, but not yet driving enterprise-wide change.

There is a long road between experimentation and transformation.

WHERE AI IS (AND ISN'T) BEING USED

Some business functions are natural fits for AI (Fig. 4.1). Marketing and sales are obvious examples. Generative AI can produce endless variations of copy and visuals, while machine-learning models can optimise lead scoring, campaign targeting, and customer retention.

Customer service is another early adopter. Chatbots now handle a significant share of customer interactions. They're fast, tireless and, increasingly, effective. In operations, AI monitors networks, detects anomalies, forecasts demand and manages logistics.

But not every function is moving at the same pace.

In manufacturing, AI has shown promise for years, particularly in quality control. Machine vision systems can detect tiny defects on production lines that human inspectors might miss. But

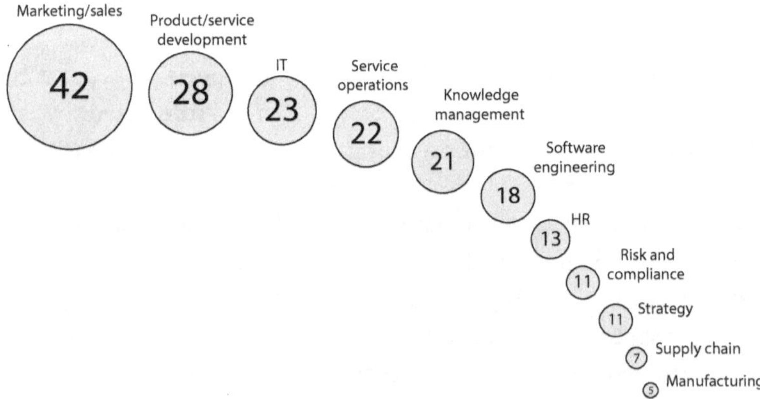

Fig. 4.1. Percentage of generative AI penetration by business function (adapted from 2025 McKinsey survey).[20]

adoption at a larger scale remains limited. Manufacturing is complex, expensive to digitise and often lacking in the large, labelled datasets AI needs in order to perform well.

Strategy and risk are even more tentative. AI is well suited to analysing risk, but only if the right data are collected, and many companies haven't made that investment. As for strategy, many leaders are reluctant to hand over the reins to a machine. Understandably so. While AI can support analysis and drafting, strategy is still deeply human, reliant on judgement, negotiation and alignment.

HR, too, is just starting to find its footing. Generative AI can help with administrative tasks, like writing policies, screening CVs and answering internal queries, but more complex (and potentially higher pay-off) developments lag behind. AI coaches, for example, are starting to appear, offering low-cost, always-on support for leadership development. Whether they'll be truly useful or just digital gimmicks remains to be seen.

FOUR LEVELS OF AI USE

AI is impacting a range of business functions across a range of sectors. Business leaders are usually keen to ask me which use cases are right for their business. This can be a difficult question to answer. The pervasiveness of AI often means that there are *too many* possible ways of applying AI. Organisations need to look for high-value use cases that offer clear and significant return on investment.

To help make smarter decisions about where to apply AI, I find it useful to think about use cases in different categories, based on complexity, generality, ease of implementation and return on investment. Figure 4.2 presents four levels of use cases.

Level one use cases are the most general, easiest to implement and require the least investment. As an example, you could

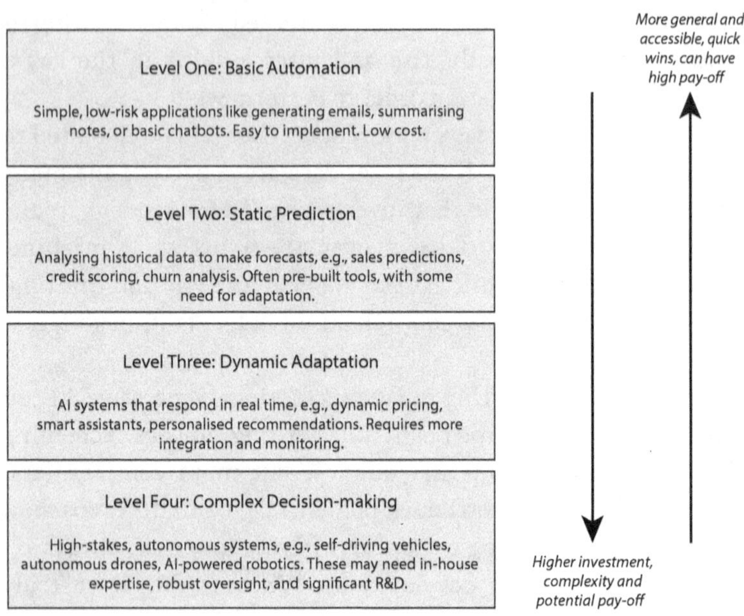

Fig. 4.2. Four AI use case levels.

simply make ChatGPT (or other generative AI tools) available to employees for basic administrative tasks. This requires minimal investment and no implementation effort beyond providing ChatGPT access.

Level two use cases require more investment but have the potential for greater return. For instance, you could develop a bespoke AI chatbot for your organisation, trained on your internal documents, so that staff can more easily access information on policies and procedures. This is relatively straightforward but does require training on internal documents, which may require some restructuring or rewriting of documents to make them suitable.

At level three are use cases with a degree of complexity, typically more specialist in nature but offering very high returns on investment. As an example, imagine a recommender system that drives customer retention by offering personalised discounts based on a customer's purchasing history. This solution requires a mature approach to data collection and management and a specialist AI model trained to predict customer behaviours. It also requires you to consider data privacy, what uses of their data customers have agreed to, and how customers will perceive such an application.

The most complex AI applications reside at level four and require the invention of new AI algorithms, architectures and approaches, necessitating in-house research and development (R&D) capability. For example, autonomous vehicles demand real-time AI processing, decision-making and rigorous safety compliance at scale. It is unlikely you will try to build a self-driving car without significant in-house expertise.

It is important to match your ambitions with your capabilities. AI maturity models can help assess where your organisation stands. Gartner's model, for instance, outlines five levels of maturity, ranging from 'Awareness', where an organisation has an interest in AI but hasn't implemented anything, all the way to 'Transformation', where AI is part of a company's DNA.[21] Similarly, MITRE's model has five levels of maturity and assesses these against six pillars needed for AI: strategy, data, organisation, technology enablers, responsible use and monitoring.[22]

If your organisation's AI maturity is at an early stage, focus on level one or level two use cases. Build confidence. Show value. Learn. If you're more advanced, you might target adaptive systems or even breakthrough innovations. But don't assume maturity just because you're enthusiastic. Be honest about where you are and build from there.

WHEN NOT TO USE AI

This chapter has focused on where organisations are currently using AI and has provided guidance on which AI use cases you might wish to adopt. However, it is worth pointing out that AI isn't always the answer. AI shouldn't be adopted for the sake of it, or because of fear of missing out. You need to think systematically about where AI makes sense in your business and not rush to adopt it just because others are, or because a vendor says you must. I'll get into the whys and hows of adopting AI in later chapters, but for now it's important to note that there are some situations where I would not recommend using AI at all. These situations relate to the reliability of the technology, the readiness of your organisation, and the social and ethical implications.

Reliability of AI

Remember that AI is just a large-scale statistical pattern matching engine. That means AI sometimes gets things wrong. Even the best AI will still make mistakes. Errors can come in different forms. In popular generative AI chatbots, errors are known as hallucinations. These can be relatively benign, and even fun. When Google's Gemini came out, I couldn't resist asking it, 'Who is Jon Whittle?' It came back with two sentences. First: *Jon Whittle is a world-renowned expert in software engineering and the human aspects of technology.* Wow, I thought. This thing's good! Then: *Jon Whittle is also a Fellow of the Australian Academy of Sciences.* I'm not a Fellow of the Australian Academy of Sciences. So that got me thinking: this thing's *really* good. It can tell the future!

This chatbot hallucination was harmless and it made me chuckle. But even for generative AI, some errors can be more serious. In 2024, Air Canada lost a tribunal case in which its customer chatbot incorrectly offered a passenger discount, only for the firm to renege on that promise, saying the chatbot had invented the discount.[23] Air Canada was forced to pay out anyway.

Beyond chatbots, the accuracy required of AI depends on the application. Using AI to recommend movies to watch on Netflix is relatively risk free. No one dies from watching a bad movie. But an AI that takes over from a doctor in making a medical diagnosis could have very serious consequences, and I would expect very high levels of accuracy from the AI as well as human oversight in those cases.

Ultimately, leaders should be aware that because of its statistical nature, AI is not infallible. Neither are humans, of course. Human decision-making is highly dependent on a range of factors such as fatigue, emotional state, personal bias and even hunger. In fact, this is often touted as one of the advantages of AI: it doesn't need to sleep, it won't take a break for lunch and it won't let a fight with its girlfriend or boyfriend cloud its choices.

But business reputation can be more badly affected by a wrong AI decision than a human one, so it behoves you as a leader to assess what level of accuracy you need from AI and whether it can deliver.

Readiness of your organisation

AI isn't an off-the-shelf technology. Whilst your employees can get started immediately with publicly available generative AI chatbots, the real value of AI comes when you develop AI solutions tailored to your context. And this requires investment of time and resources. Depending on the AI application, you will need access to high-quality data in a form and at a scale that you can use it to train your AI system. Many organisations have poor data management practices, with data scattered across the business, not easily accessible or shareable, and perhaps even not conforming to privacy laws. Even the most progressive AI companies spend a lot of time cleaning up their data and making it AI ready. Telstra, Australia's largest mobile network, known for its proactive approach to AI, is having to reduce the number of its data platforms from 80 to three to accommodate its AI ambitions.[24] That can take years of work.

Another aspect of AI readiness is whether your organisation has AI skills or can partner with another business to get them. Some companies rely exclusively on tech vendors or IT consultants to deliver AI solutions for them, whereas others have an in-house team. In general, the latter is probably better, although it may require greater investment and will present challenges in hiring AI experts given market demand. Relying solely on an outside contractor presents risks because it creates a dependency on an external organisation.

Finally, culture is all-important when it comes to AI readiness. If your employees are not ready to embrace AI, it is unlikely to be successful. Culture extends to your stakeholders and customers as well. How will they react to an increased use of AI? If there is likely to be significant resistance to your AI application, it might be time to rethink whether the benefits outweigh the potential fallout.

Social and ethical implications

It goes without saying that there are some things you can do with AI that you simply should not do. Before asking the question, 'Can AI do this?' stop and ask, '*Should* AI do this?' AI applications may be prohibited by law in some jurisdictions. The European Union's AI Act, for example, bans AI for social scoring, scraping for facial recognition systems and using AI to exploit a person's vulnerabilities. Beyond the law, however, countries have moral norms that customers normally expect to see followed. In the 2016 US Presidential campaign, Cambridge Analytica sold voters' data scraped from Facebook accounts to campaign offices. Even where Cambridge Analytica's actions were technically legal, they were certainly unethical,[25] and this lack of attention to ethics resulted in a major backlash.

Organisations sometimes develop AI solutions in good faith but haven't fully considered public perception. Australian home improvement retailer Bunnings was caught out when it used facial recognition systems in its stores that collected personal data of customers without their consent. Their justification was that the system protected staff against a string of violent incidents and organised crime in stores, but that didn't a prevent a media firestorm, forcing Bunnings to retire the technology.[26]

The potential negative societal implications of AI are vast, ranging from impact on jobs, to environmental impact, and to discrimination. It might be tempting for an organisation to use AI to supercharge its business, but you need to ask whether the application is aligned with your corporate values and wider societal expectations.

If in doubt, think harder.

QUESTIONS

- How is your organisation currently using AI? Make a list of as many applications as you can and match them against the four levels in Fig. 4.2.
- In what business functions do you think your organisation is likely to get the highest return on AI investment?
- Can you think of any AI applications in your business that you would say no to?

Part II
Why AI?

Part II of this book covers reasons why you might be considering adopting AI.

When I talk to companies about AI, the question at the top of their minds is always, 'Where do I start?' Before answering this question, though, there's an even more important question that should be addressed: why start at all? As a leader, you will have some sense of the benefits of applying AI in your organisation. But you may not have clearly articulated exactly which of those benefits will apply in your business and how. What specific benefit are you hoping to achieve? Which part of your business are you hoping to achieve it in? Without concrete answers to the 'why' question, it may be difficult to convince your organisation to go with you on the AI journey.

There are many purported benefits of AI. It is important that you go beyond high-level goals like 'AI will improve our productivity' to really understand where, how and, indeed, if, those goals will be achieved from AI.

Each chapter in Part II covers a potential benefit of AI, giving you a balanced account of whether that benefit can be achieved in practice.

- *Productivity.* One of the most common reasons for applying AI is automation of processes and tasks. But how easy is this to achieve in practice? Where can AI improve productivity and where might it add extra work?
- *Cost reduction.* AI can help with your bottom line by streamlining, or even eliminating, costly parts of your business. But AI requires both up front and ongoing investment as well. How much will you have to invest and what will your return be? And what will be the impact on your labour budget?
- *Adding customer value.* Automating your processes can lead to an improved customer experience, but beyond automation, how can AI transform the way you interact with customers, lead to better customer retention and ultimately to happier customers?

- ***Disruption.*** **Doing the things you are already doing faster and cheaper is one thing. Doing completely new things is something else altogether. In many cases, AI could completely disrupt your business model giving you massive competitive advantage.**

5

Will AI improve productivity?

THE MOST CITED BENEFIT OF AI is that it will 'supercharge' productivity (Fig. 5.1). In April 2023, Goldman Sachs predicted that generative AI will bring 'sweeping changes to the economy', claiming a 1.5 per cent increase in productivity annually over the

Fig. 5.1. AI supercharges productivity (image generated by ChatGPT).

next decade and a 7 per cent rise in global gross domestic product (GDP).[27] In the same year, a McKinsey report said that 'generative AI is poised to unleash the next wave of productivity',[28] arguing that AI will automate 50 per cent of all work activities.

We should be cautious about AI predictions. Since its inception, the field has been obsessed with crystal balling. But crystal balls get things wrong. In 2016, Nobel Prize-winning AI guru, Geoffrey Hinton, said that AI would work better than radiologists within five years and we should stop training human radiologists.[29] But the demand for human radiologists is higher than ever. Over 10 years ago, a famous study by Frey and Osborne from Oxford University forecast that 47 per cent of jobs would be automated.[30] This prediction has since been debunked.

The problem with predictions is that they're hard to get right, especially when they're about the future! The *Pessimists' Archive*[31] is an archive of past technology predictions. Some of these are ridiculous in hindsight, such as the 1909 article claiming that aeroplanes were just a fad. But some predictions are eerie when read years later, like the 1990 newspaper clipping that claimed the Sony Walkman was increasing social isolation as a generation walked the streets lost in their own world.

When it comes to AI, predictions are everywhere and always have been. In the 1950s, AI pioneers thought human-like intelligence was just a few months away. On the other hand, almost everyone in the 1980s thought neural networks were a passing trend. Daniel Shrader reflects on past AI prophecies and concludes that they are mostly rubbish.[32] I tend to agree. In fact, the situation has become so bad that I only ever give one prediction about AI: every AI prediction is wrong!

If we can't trust predictions about the future of AI productivity impacts, how can we begin to understand if AI will make us more productive? The only realistic course of action is to look to science and examine the evidence of how AI is impacting productivity *now*, acknowledging, of course, that past performance is not an indicator of what may eventuate in the future.

WHAT IS PRODUCTIVITY?

Before delving into the evidence on AI and productivity, I ought to define what is commonly meant by productivity, and why it is important.

For the purposes of this book, I will limit the discussion to labour productivity; that is, how efficiently employee work hours are transformed into outputs. I will also make a distinction between individual, organisational and national productivity.

Individual productivity is how efficiently an individual manages their time and resources to complete tasks. Individual productivity can be measured using a variety of metrics, which typically depend on the task itself. Some worker tasks may be easy to measure, such as how quickly a customer service representative resolves a query. Some tasks are harder to measure – for example, how would you measure the impact of mentoring on a colleague? Aggregating all a worker's tasks into a single metric is typically very difficult unless the worker has a very repetitive occupation.

Organisational productivity is how well an organisation achieves its goals. In some sense, corporate key performance indicators (KPIs) are a good measure of organisational productivity. For example, a product company may have a KPI of the number of new users divided by number of sales staff. Or a research organisation could measure the number of high-quality research papers produced versus number of research staff. In principle, one could measure every individual's productivity and then aggregate that at an organisational level. However, this can be complex in practice.

National productivity is the economic efficiency of a nation. There are many ways to measure national productivity, but one of the simplest is to measure labour productivity as the GDP per hour worked.

Historically, productivity has been considered important because of the link between productivity and 'success', however that is defined. For individuals, productivity and happiness are correlated.[33] For organisations, productivity impacts several factors, including customer satisfaction and competitiveness in the market. Nationally, increasing productivity leads to better standards of living.

AI AND PRODUCTIVITY

AI can impact all three types of productivity. As an individual, maybe I can write emails faster with AI. As an organisation, maybe my teams can resolve problems more quickly if they use AI. Nationally, AI could speed up manufacturing, giving a country a competitive edge economically.

But what does the evidence say about AI's impact on productivity? Is it wholly positive?

At the risk of giving away the punchline, AI can certainly improve productivity on all three levels; however, productivity improvements are not a given. Productivity is a complex phenomenon and it is easy to introduce efficiencies in one area but make things worse in another. We will see why below.

AI AND INDIVIDUAL PRODUCTIVITY

Studies on AI productivity are still emerging. It is a fair summary to say that on individual tasks, AI has a good chance of increasing productivity, but it isn't always the case. It really depends on the task and how AI is adopted.

In 2020–2021, a Fortune 500 software company put into production an early generative AI system used by 5,200 customer support agents.[34] The system was used in an advisory mode, providing real-time recommendations for agents on how to respond to customers. This is an example of AI being used within a business for a specific, individual task. The results showed a 14 per cent increase in productivity on this individual task, measured as the number of issues resolved per hour. For less experienced agents, productivity increased by 35 per cent.

A more recent Harvard-led study[35] in 2023 examined GPT-4 use with over 750 consultants from the Boston Consulting Group. Consultants were asked to perform a set of tasks representing typical job activities, such as ideating a new product, analysing data and writing persuasively. In contrast to the above study, where the tasks were part of the agents' actual jobs, this study assigned tasks in an experimental setting over a five-hour period; however, the researchers took steps to ensure the tasks were realistic. The results showed that consultants were 18 per cent faster on easier tasks when using generative AI, and the quality of the answers was also higher. On harder tasks, such as developing business strategy, consultants using AI still completed the tasks more quickly. However, the correctness of their answers degraded by 19 per cent.

At time of writing, the most recent real-world study of AI and productivity involved 776 experienced professionals at Proctor & Gamble.[36] As well as some other reported benefits of AI, the study showed that individuals randomly assigned to use AI performed as well as a team of two without AI.

The task of computer programming is often considered a 'killer app' for generative AI. AI's ability to generate code partially automates one of society's most time-consuming and high skills-demanding jobs. It has been estimated that software contributes US$1.1 trillion to the US economy, so it is clear that an increase in individual productivity in software coders has the potential to boost national productivity.

Microsoft Research carried out a controlled experiment with GitHub Copilot, an AI coding assistant.[37] Software developers were asked to implement an HTTP server in JavaScript as quickly as possible. Those using AI completed the task 55 per cent faster than those not using AI. These productivity improvements are backed up by other studies. A real-world randomised controlled trial with almost 5,000 developers at a Fortune 100 company revealed a 26 per cent productivity improvement.[38] Other studies show similar gains, typically ranging up to 50 per cent.[39] And surveys show that up to 90 per cent of developers self-report productivity improvements.[40]

But AI does not always increase individual productivity.

A survey of 2,500 professionals by Upwork found that generative AI increased workload for 77 per cent of workers.[41] Forty-seven per cent said they didn't know how to unlock productivity benefits. The study points to barriers such as the need to verify and/or correct AI outputs, the need for AI upskilling and unreasonable expectations about what AI can do.

Intel carried out a study with 6,000 users to see if their AI personal computers would save time on common knowledge work tasks such as writing emails and transcribing meetings but found that users spent longer on these tasks because of the need to prompt the AI in the right way.[42]

A recent CSIRO study examined the daily use of Microsoft 365 Copilot by 300 employees of a government organisation.[43] Whilst the majority self-reported productivity benefits, a sizeable minority (30 per cent) did not. And use of Copilot was limited to a small number of common tasks, like writing an email or summarising meeting notes. Participants struggled to find benefits from Copilot in more creative tasks. Even those workers who reported productivity improvements expected greater productivity benefits than were delivered. A similar study, carried out in 50 Australian government agencies, found similar results, concluding that effective AI training is a prerequisite for productivity gains.[44]

Studies have consistently shown that less skilled professionals benefit more from AI than the more skilled. The customer support agent study, for example, found that experienced agents were less likely to rely on AI advice, ignoring 65 per cent of its recommendations.

There can be significant downsides to AI in computer programming. One study found that 52 per cent of programs generated by ChatGPT contained errors, and 77 per cent were verbose.[45] This could lead to major productivity issues if the effort needed to test generated code increases.

On balance, whilst productivity increases from AI are far from universal, it does appear that individual tasks can benefit. It depends on the nature of the task, of course, but it also depends on the AI solution and how it is introduced.

Top tip: Ingredients for productivity success

A classic study of AI and productivity by Brynjolfsson and colleagues showed a 14 per cent productivity increase for customer support agents assisted by AI.[35] Beyond this simple number, however, what factors led to success?

- A very well-defined task with well-established metrics that can be used to measure success.
- Access to plenty of labelled historical data – in this case, customer interactions that had been recorded over a long period of time, both with successful and unsuccessful resolutions.
- It wasn't a case of just applying an off-the-shelf system. Specialist machine-learning algorithms were used to fine-tune an off-the-shelf system, training on organisation-specific

data. A significant amount of investment was required to realise the benefits fully.

- AI was designed to assist, not replace, humans.

If you look beyond the headline, you see that AI productivity increases aren't guaranteed. They are achievable but you need to think about the quality of data, the costs of implementation, the fact that some workers benefit and some do not, that tasks need to be well defined and that having humans work alongside AI may be a good bet.

AI AND ORGANISATIONAL PRODUCTIVITY

The OECD has estimated productivity benefits of traditional AI to be zero to 11 per cent at the organisational level.[46] Traditional AI – machine-learning applied for an industry-specific task – remains the most widely used form of AI in industry. For decades, traditional AI has been used in factories, laboratories, transportation and hospitals.

Independent studies in Germany, Italy and Taiwan also show increases in organisational productivity, [47] but two studies in the United States found no correlation between AI and productivity when removing the effect of related IT technologies (cloud computing, computer hardware, general IT systems).[48] A study in Ireland[48] could not show a significant association between AI use and productivity once other aspects (e.g. skills, country of ownership) were factored out.

Measuring productivity at the organisational level is notoriously difficult. Unless you roll out time-and-motion studies, you're unlikely to get an accurate measure of how individual employee productivity aggregates to an organisational level. From an AI perspective, it's difficult, if not impossible, to attribute changes in an organisation's productivity to the introduction of

AI. Businesses are sensitive to many social and organisational factors, any one of which could be the reason for a change in productivity, making it hard to say if AI made the difference or not. For example, you might introduce AI and see a productivity uptick. But suppose the introduction of AI coincided with a new hiring round. Did AI lead to the change or was it down to a bunch of smart new employees who just happened to be hired around the same time? Correlation, as they say, does not equal causation.

There have been many examples where the introduction of AI had unintended consequences leading to major productivity problems. Amazon's Just Walk Out grocery stores, intended to reduce labour as customer purchases would be fully automated, instead resulted in the hiring of 1,000 remote workers to verify purchases as the AI consistently made mistakes.[48] In 2021, the idea that machine learning could accurately diagnose COVID-19 from scans was debunked when 62 studies using AI were found to be methodologically flawed.[49]

When it comes to generative AI, we don't yet have reliable data on organisational level productivity improvements. Given that generative AI adoption is still emerging, it is too early to predict the impact on organisational productivity. Other than anecdotes, there really isn't any evidence that generative AI 'supercharges' an organisation. We know, however, that it takes time for a new technology to be fully leveraged, so we may see stronger correlations in the future.

Top tip: The complexity of digital transformation

It is very hard to guarantee that AI will transform the productivity of an organisation. There are too many social and organisational factors, meaning that gains in one part of your business may not be realised elsewhere. This has been known for a long time.

A seminal study by Jonathan Grudin in 1988[50] uncovered the key factors regarding why technology adoption in organisations can fail:

- It can save effort for one group of users but create additional work for others. The typical example is where an IT system is procured by a manager that makes their job easier but others' jobs harder.
- Adoption methods that worked for yesterday's technology may not work for tomorrow's innovations.
- Difficulties in evaluating the return on investment (ROI) of the tech. 'Transformative' technologies are complex and touch many parts of an organisation. This means evaluation is resistant to simple metrics/KPIs and won't fully capture where the technology is working and where it isn't.

Almost 40 years later, Grudin's insights are still relevant today. Do you recognise these same factors in your organisation's attempts to adopt AI?

AI AND NATIONAL PRODUCTIVITY

The picture at a national level is even murkier. Clearly AI hasn't yet impacted national productivity. Most developed nations are languishing with record low productivity growth. In Australia, for example, labour productivity growth is at a 60 year low.[51] In the United States, while growth has been accelerating recently, it still sits well below where it was in the early 2000s. In the European Union, labour productivity is declining.[52]

It can be argued that technology developments like AI take time to affect national productivity because companies need to figure out how to use the technology and how to put the necessary infrastructure and skills in place. This is not guaranteed, however. Experts differ on the role of digital technologies in improving productivity. Whilst there is consensus that the internet did lead to productivity improvements, the effects of mobile phones and social media are more contested, their impacts being limited to certain industries, like entertainment.[53] It remains to be seen what the effect of AI will be.

Top tip: Beware Solow's paradox

In 1987 American economist Robert Solow famously quipped, 'You can see the computer age everywhere but in the productivity statistics.' Solow's paradox,[54] also called the productivity paradox, remains visible today. Sometimes more investment in IT in business processes can lead to worker productivity going down instead of up.

FACTORS AFFECTING AI AND PRODUCTIVITY

It is clear that AI has the *potential* to improve productivity at individual, organisational and national levels, but in each case, there are many factors at play that determine whether productivity improves or not. Success is not guaranteed.

Successful technology adoption is as much about the social and organisational aspects surrounding the technology as it is about the technology itself. Improving productivity isn't as simple as simply automating a time-consuming task. Technology adoption can have unforeseen negative consequences where you least expect them, meaning that productivity gains you thought you had secured quickly evaporate.

As a business leader tasked with introducing AI, it is important to think holistically. A growing body of research reveals that while AI has tremendous potential to enhance productivity, its real-world implementation often yields mixed results, sometimes even reducing productivity, particularly in the short term.

While AI has the potential to improve productivity, it often causes labour market disruption without the immediate compensation of new, productive roles for displaced workers. This displacement can lead to productivity losses if the workforce is not reallocated effectively.

Beyond labour issues, organisational and cognitive challenges further complicate AI's promise. Poorly integrated AI systems often increase workflow complexity, strain legacy systems and inadvertently slow down operations. Additionally, in high-skill environments, AI can impose cognitive overload on workers who must interpret machine outputs alongside their usual tasks, leading to mental fatigue and reduced productivity. These factors contribute to what researchers term the *AI productivity paradox*[55] – a disconnect between the rapid progress of AI technologies and the stagnant or declining productivity figures in many industries (remember Solow's paradox?). Cultural resistance, misalignment between AI tools and business needs, and diminishing returns after initial AI adoption further explain why AI, despite its hype, doesn't always translate into measurable gains.

Organisational Factors	
Does your organisation have a track record of successful digital transformation?	✓ or X
Can you 'stay the course' given that benefits might take time to materialise?	✓ or X
Early wins might fade. Do you have a plan for diminishing returns from AI?	✓ or X
Are you taking a system-wide approach, so that benefits in one part of the organisation aren't displaced by new problems in others?	✓ or X
People Factors	
Does your workforce have the skills to exploit AI productivity benefits?	✓ or X
Do you have the right training programs in place?	✓ or X
What happens to displaced workers – do you have a plan to reallocate and reskill effectively?	✓ or X
Are you rewarding workers for increasing productivity or just increasing expectations on them?	✓ or X
Do you have a plan to manage cultural resistance?	✓ or X
Technology Factors	
Are you already a mature adopter of digital technologies?	✓ or X
Have you fully thought through how to integrate AI with legacy systems?	✓ or X
Do you have a thorough understanding of how AI systems work in practice?	✓ or X

Fig. 5.2. Checklist of factors affecting AI and productivity.

Figure 5.2 highlights some of the major factors affecting AI and productivity. As an organisation embarking on an AI transformation, it's important to consider these and, wherever possible, introduce mitigations that aim to maximise the productivity benefits and minimise areas where productivity might suffer.

QUESTIONS

- Think about where AI might improve your own personal productivity. First, spend one week tracking what you spend your time on. Now, think about which tasks AI could automate. Does AI target your most time-consuming tasks?
- How ready is your organisation to realise AI productivity benefits? Make an initial assessment using the checklist in Fig. 5.2.

6

What does AI cost?

IN JANUARY 2025, DONALD TRUMP began his second term as President of the United States. He wasted no time setting about his agenda, signing a record 54 executive orders in his first two weeks in office. On day one of his presidency, Trump rescinded President Biden's AI executive order, which had focused on holding tech companies accountable for its safe development. On day three, Trump announced what he called 'the largest AI infrastructure project in history', a US$500 billion joint venture between Softbank, Oracle and OpenAI, to build Stargate, a network of data centres across the United States.

The announcement was meant to be America's first salvo in an AI arms race against the rest of the world. It was supposed to be the AI equivalent of a 'first strike' that would knock out opponents in Europe and Asia.

And then, just one week after Trump became President, China fired back.

On 27 January 2025, billions of dollars were wiped off the US stock market in 24 hours after Chinese startup DeepSeek released its R1 chatbot. R1 quickly became the number one app downloaded on the Apple Store, leapfrogging ChatGPT, which, until then, had managed to fight off all newcomers. R1's performance was at least as good as ChatGPT's but, more than that, there were advances in R1 that no one saw coming.

First, DeepSeek R1 was a specialist in AI 'reasoning'. It excelled at tasks that require breaking down a problem into step-by-step solutions. And it could explain that reasoning process to the user.

Second, DeepSeek R1 was released as open source, meaning that researchers anywhere could replicate and build upon it. This contrasts with the models from US tech companies that closely guard proprietary information.

These two innovations by themselves are enough to grab media headlines, but what made commentators see R1 as a real threat was the cost of developing it. By introducing hardware and software efficiencies, DeepSeek somehow managed to produce an AI just as capable as the best GPT models but that was 96 per cent cheaper to use.[56] The initial headlines focused on a figure that DeepSeek released saying R1 cost only US$5.5 million to train; that's compared to billions of dollars in training costs from more established companies. It was this figure that caused the US stock market to crash. While it quickly surfaced that DeepSeek spent considerably more than US$5.5 million, R1 was still significantly cheaper than competitors, and the contrast with the Stargate figure is stark.

DeepSeek was seen as a major threat to US dominance in the AI field. Not only was R1 cheaper, but any data entered into R1 via a web browser or its app ended up on Chinese servers. Several countries quickly moved to ban R1 on official government devices.

In the race to develop cost-effective AI, 27 January 2025 marks a turning point. It should have been obvious even before that date that money is a major enabler of the AI industry, but if anyone was unsure, there was no doubt after January 27 that when it comes to AI, cost is king.

RETURN ON INVESTMENT

One of the key drivers for organisations when adopting AI is cost savings. AI has proven for decades that *if implemented well* it can lead to substantial savings. This phrase is important, though. Many AI projects fail – by some estimates up to 80 per cent.[4] If you succeed with AI, you should reap the financial benefits. But success isn't guaranteed.

To give a sense of the potential financial value from adopting AI, Table 6.1 collects some reported AI success stories and the positive impact on the bottom line. Take this list with a pinch of salt, though. It turns out to be surprisingly difficult to find rigorous data on AI cost savings because much of the data is proprietary and/or companies are not good at estimating return on investment.

We also don't tend to hear about the failures of AI adoption apart from a few isolated cases in the media. The simple truth is that AI has the *potential* for significant cost reductions, but there's also a serious risk that AI investment will spin out of control, leading to abandoned costly AI projects.

AI can fail for several reasons (Table 6.2). Sometimes the technology just doesn't work very well. McDonald's abandoned an AI drive-through ordering system after the program had difficulty understanding customers' requests, in one case adding 260 chicken nuggets to a single order![57] Other reasons for failure include unanticipated customer backlash. Science magazine *Cosmos* drew criticism from journalists after using AI to generate articles.[58] (The publisher's response explained its use of AI was a controlled experiment designed to assess the technology.) Some companies have been accused of violating ethical norms or discrimination laws in their use of AI. iTutorGroup was forced to pay US$365,000 after using an AI recruitment tool which was later shown to discriminate against older women.[58]

Table 6.1. Some reported examples of AI cost savings.

Company	Use of AI	Reported savings
Comcast	Implemented an AI tool named 'Ask Me Anything' to assist customer service agents, resulting in a 10% reduction in time spent per conversation	Comcast report authors claim millions of dollars in savings due to less time spent resolving customer queries[59]
The Adelaide Score	AI system that predicts when general surgery patients are ready to be discharged from hospital, thus freeing up beds	A trial at the Lyell McEwin hospital reported savings of AU$750,000 over the four-week trial[60]
Integrated Energy Systems	Research on neural networks to optimise energy forecasting by learning from historical forecasting errors	Authors claim 63.5% cost reduction across several case studies compared to traditional energy forecasting methods[61]
NASA Turbofan engines	AI-assisted predictive maintenance of aircraft engines	Experiments show 13% cost savings compared to traditional maintenance approaches[62]

Table 6.2. Some reported examples of AI cost blow outs.

Company	Use of AI	Reported costs
Zillow	Real estate marketplace company created 'Zillow Offers' using a machine-learning algorithm to make cash offers on property with flip potential	Error rates of the algorithm were as high as 6.9% leading to a US$304 million inventory write-down[58]
Banks and financial services	A study of five US banks and financial services firms uncovered a range of 'AI incidents' causing unintended financial losses: bias in mortgage approval algorithms, calculation errors, deepfake social media accounts, etc.	21% short-term losses in stock market values[63]
Los Angeles Unified School District	Implemented an AI chatbot named 'Ed' for student assistance, which faced technical failures and data privacy concerns	Project shelved after three months resulting in US$3 million in losses[64]
Tyndaris Investments	AI-assisted trading	One investor sued for US$20 million in losses[65]

IMPACT ON JOBS

An issue related to AI cost savings is the impact of AI on jobs. The public's perception of how AI will affect jobs can influence government policies and organisational strategy.

Economists are divided when it comes to AI and jobs. Much of the literature on the topic takes a forward-looking view, trying to predict jobs displacement in the coming decades. Such predictions are notoriously unreliable. I find it's more instructive to take a historical view. A Centre for Research on Economic Policy (CEPR) study found that industrial robotics has led to a small but significant increase in jobs.[66] Many commentators argue that automation has led to more jobs because of increased opportunity. On the other hand, one study found that each robot introduced into an industrial process has led to the loss of five jobs.[67]

It is very difficult to know where the truth lies. Although automation may clearly lead to job losses in some areas, understanding the indirect impact on new job opportunities is much harder. A loose consensus appears to be that the nature of jobs will change due to AI. While overall employment might improve, there will be winners and losers as traditional jobs evaporate, and new professions appear. For a socially responsible organisation introducing AI, the onus lies with that organisation to consider retraining opportunities for displaced staff.

From an organisational point of view, a natural question to ask is whether AI will enable savings in the labour budget.

Table 6.3 provides some examples where companies have publicly stated that AI led to reductions in workforce numbers. Such reports should be interpreted with care. Sometimes companies attribute job losses to technology as a 'decoy' to avoid discussing internal challenges. However, the examples in Table 6.3 are all from companies proudly presenting themselves as early adopters of AI and presenting the case to shareholders.

While AI adoption often reduces labour costs, it can sometimes lead to increased labour budgets. This may be due to the need for specialised personnel to develop, implement and manage AI systems, as well as the need for training. The US Bureau of Labour Statistics projects a 26% growth in computer and information research scientist roles from 2023 to 2033.[68] But AI talent is expensive. Some sources report that US salaries for an AI engineer have risen almost 50 per cent since 2021.[69] For senior AI executives, salaries can be out of reach for many companies, meaning that only a few can afford to execute a thorough AI transformation.

An alternative to hiring new talent is to train existing employees. The World Economic Forum estimates that by 2025,

Table 6.3. Reported examples of AI leading to workforce reductions.

Co.	Headline	Story
Sky	Sky to cut 2,000 call centre jobs amid AI shift, Broadband TV News, 28 March 2025[72]	Reported culling of 2,000 customer call centre jobs due to increased digitisation in customer service chatbots and other tools
Ocado	Ocado to cut 500 technology and finance jobs as AI reduces costs, Butler, *The Guardian*, 27 February 2025[73]	Reportedly, AI led to productivity increases in engineering teams. Coupled with new robotic packing devices, Ocado announced job cuts
Workday	Workday erases 8.5% of workforce because of AI, Kunert, *The Register*, 5 February 2025[74]	1,750 job losses due to the company's decision to pivot towards AI-focused use cases
BT	Companies that have already replaced workers with AI in 2024 & 2025, O'Sullivan, Tech.co, 16 May 2025[75]	BT plans to replace 10,000 jobs with AI by the end of the decade

50% of all employees will require reskilling due to AI and automation adoption.[70] Research from the Federal Reserve Bank of New York indicates that firms planning to adopt AI are more likely to retrain their current workforce rather than reduce headcounts, implying an increase in labour expenditures for training and development.[71]

COSTS OF ADOPTING AI

AI adoption does not come free of charge. Indeed, the investment required to adopt AI fully across an organisation may be significant.[76] As a leader tasked with driving AI adoption, you should think carefully about the investment required. This typically falls into the following categories.

Computing infrastructure

AI needs infrastructure to run. This includes both hardware and software. For hardware, you have a choice: run AI on your own computing infrastructure or rely on cloud services. Both incur costs. In the former case, you will need to budget for servers, graphics processing units (GPUs) and any other computing devices needed to run your AI. In the latter case, there will be costs for cloud-based services, data storage and compute power. For software, you will need to invest in AI platforms, machine-learning frameworks, analytics tools and automation software. Whilst some software is free or open source, you may also need to pay for proprietary software.

Data

AI relies heavily on data – lots of data. You will incur costs associated with data collection, whether that means integrating internal datasets or purchasing external data. The key to getting AI to work properly is good data hygiene, so you will need to factor in costs of 'data cleaning' to ensure your data is high quality, consistent, and free from bias. You will also need to pay careful attention to data privacy laws and there may be additional compliance activities to factor into your budget.

Development and integration

Once you have the data, you will need to develop, test and refine AI models – either models developed in-house or adapted from elsewhere. It is important to understand that AI systems don't run in isolation from the rest of your IT infrastructure. In fact, to be effective, AI needs to be fully integrated with other systems. This will incur integration costs.

Operations and maintenance

Any IT system incurs operating costs. You will need to monitor AI closely for accuracy and performance and to head off any reputational risks associated with incorrect or problematic responses.

Talent and training

You will require a range of skills to implement AI properly, ranging from software developers, AI professionals, governance experts and project managers. Depending on whether you build an internal team or outsource, you will incur costs of hiring, training and upskilling, or consultancy.

Compliance and governance

There are an increasing number of laws and international standards that govern AI. These range from privacy laws and industry-specific laws to general AI laws such as the EU AI Act. Ensuring compliance with these laws can be costly and will require changes to your governance structures, risk management processes and reporting mechanisms.

Change management and communication

The use of AI may be a significant departure for your staff and your customers. To prepare your workforce for AI adoption, you will need strong change management practices to avoid resistance to change in workplace practices and, in some cases, organisational structure. You will need to ensure effective communication with your customers, especially in sectors where there is public distrust of AI systems.

Risk mitigation and cybersecurity

AI brings new cybersecurity vulnerabilities which you will need to manage. AI is never 100 per cent accurate so it's a good idea to think about building in redundancy and back-up systems in case customers are unable or unwilling to use AI.

Top tip: Know what questions to ask

As a leader, you are not expected to do the detailed budget calculations to figure out exactly what AI will cost your organisation (unless you happen to be the CFO), but you are expected to know the right questions to ask when you're presented with a budget. If you're struggling with this, consider using generative AI to help. Try using the following prompt:

> *I'm interested in the cost of adopting AI in my organisation. Give me a list of questions, organised by category, that I should ask the Chief Financial Officer/Chief Information Officer (CFO/CIO).*

A more refined prompt could be tailored to your organisational context. If you are using an enterprise generative AI tool with appropriate privacy features, you can also consider uploading a draft budget and have AI ask questions directly of the budget.

COSTS OF GENERATIVE AI

All AI deployments – whether traditional or generative – incur costs. Given the recent popularity of generative AI, though, it is instructive to spend some time unpicking costs specifically associated with this technology. Generative AI is often perceived as low cost because many tools are available to consumers for a low-price monthly subscription (or even for free). However, generative AI isn't necessarily low cost: it really depends on what you want to do with it.

It is generally quite difficult to estimate the costs of implementing generative AI in your organisation. Costs of generative AI change all the time as tech vendors evolve their pricing models. And costs depend heavily on the kind of AI application you're working on. However, there are different ways to use generative AI in your business, and whilst the specific costs will evolve over time, it is worth understanding that different levels of complexity in how you apply generative AI will result in significantly different levels of investment required.

Generative AI systems must be trained on vast quantities of data before they are released to the public. Training an AI with billions of parameters is very expensive. By some estimates, Google's Gemini system cost upwards of US$200 million to train.[77] And, notwithstanding the DeepSeek experience, the cost to train generative AI is likely to remain high.

Somone ultimately must pay the bill. Historically, the cost has been covered by investors. Amazon invested US$8 billion in AI company Anthropic, and Microsoft invested US$13 billion in OpenAI. At some point, AI companies will need to generate revenues to cover these costs, and that means passing on the costs to consumers, with most of those costs expected to fall on enterprises.

So if you are considering adopting generative AI in your organisation, exactly how much will it cost?

Off-the-shelf generative AI chatbots

The easiest way for a company to use AI is simply to allow staff to subscribe to AI tools. For general administrative tasks, you can just allow staff to subscribe to any tools they like and charge costs to a corporate account. Most generic AI tools (e.g. ChatGPT, Gemini) cost US$20–30 per month per user (for their most basic version). There are now hundreds of subscription-based AI tools that cover a variety of tasks including text and image generation, grammar checking, project management and sales. Allowing employees complete freedom can be an expensive business (see Table 6.4).

A more managed approach to generative AI is to take out an enterprise-wide subscription. Enterprise versions of popular AI tools typically come with extra features, improved security, and the ability to integrate corporate data. ChatGPT Enterprise, for instance, provides a guarantee that information uploaded by staff won't be used by OpenAI in training future versions of ChatGPT. This means you can allow ChatGPT to access sensitive data without worrying about data leakage. Enterprise AI is often a cost-effective managed solution for larger enterprises.

Custom generative AI services

Out-of-the-box generative AI is of limited use. If you want to use AI for customer-facing products, you will need to develop your own custom AI service. Common applications include customer service and e-commerce chatbots. Pricing for this kind of customised generative AI application works differently. Your service will need to use a vendor-provided API (application programming interface); that's IT speak for an interface that allows your own code to access AI without going through a public portal. In this case, you'll be charged depending on how many times you access the API and what volume of information you exchange with the API.

The costs of running your own customised AI vary enormously, depending on which AI provider you use, which of their models you use (high performance but expensive versus low performance and cheaper), and the volume of data that goes through the AI.

As an example, suppose your organisation wishes to use a large language model to analyse customer comments on your social media channels so that recurring issues can be identified. You will need a software development team to build your app. This app will access your choice of LLM through the provider's API. You can use the API in different ways: you might send the customer reviews one by one as they come in, or you might wait and send them in batches every few days. Either way, you'll incur API costs every time you use it. It can be tricky to estimate the exact costs but for a simple application like this one, assuming a relatively high volume of 30 comments per minute, a reasonable estimate is around US$3,500 per month.[78] This is only the cost of using the LLM to analyse comments and does not include staff time to develop the app or any auxiliary costs to run it.

It is important to factor in LLM costs because they are not fixed. Without proper management, you might find that the number of customer comments spikes and you are left with an unexpected bill.

An alternative to using the API of a managed provider is to use an open-source LLM. This has the advantage that you don't pay per request so can manage costs more easily, but you would need to run the LLM on your own computer infrastructure. This incurs costs associated with the computer hardware and also staff expertise to configure the LLM and run it locally on your own machines or in the cloud.

Develop your own model

The previous example used a customised approach to access a provider's LLM for a business application, but what if you want to develop your own in-house model? For example, let's suppose that your company sells specialist automotive parts. There might be limited information on these parts in the public domain, so a generic LLM can't provide very good answers if asked about them. But you could train an LLM on your own proprietary data. That way, if a customer asks a detailed question about a particular car part, your LLM can respond appropriately.

You are unlikely to develop your own LLM from scratch for this purpose – the training costs would be prohibitive. Instead, a common approach is to *fine-tune* an existing LLM. Fine-tuning means taking a specialist dataset – essentially a list of common queries and expected responses for that domain – and feeding it to an LLM so it relearns what kind of responses it should give. Fine-tuning can be done using a provider's API or using your own computer infrastructure. The computer costs are modest but there may be significant costs (e.g. labour) associated with preparing the company-specific dataset and setting up the fine-tuning process.

In summary, generative AI is often seen as an off-the-shelf technology. In practice, however, this is not the case. Business

Table 6.4. Example computer costs of different ways of using generative AI. Costs are illustrative only.

Use case	Assumptions	Approximate annual cost (US $)
Employee individual subscriptions	Medium-sized business of 1,000 employees 40% of staff have on average 2 subscriptions each US$25 per subscription	240,000
Enterprise AI	One enterprise-wide subscription for 250 employees	90,000–180,000 per subscription[80]
Custom AI application	Simple customer comment analysis API costs only	Managed provider: 42,000
Fine-tuning	Fine-tune on 100 corporate web pages Compute costs only	Negligible

applications involve costs of computation and staffing and you need to understand these costs before embarking on an AI project. You also need to monitor costs carefully as they can become significant. According to IBM, executives are now commonly cancelling generative AI projects due to spiralling costs.[79]

QUESTIONS

- What type of AI deployment is most appropriate for your organisation? Off-the-shelf AI tools or custom built? If the latter, will you develop internally or outsource?
- Have you considered all categories of investment needed to develop and deploy your AI system? Are there any indirect costs you have not included?
- What will the impact on your labour budget be? Will you achieve cost savings or will you incur a net cost due to training and upskilling needs?

7

Will AI bring value to my customers?

DESPITE HOW CLEVER AI IS, it can be dumb too.

Generative AI, for example, can solve very complex math problems, scoring over 90 per cent on maths tests[81] and surpassing PhD-level intelligence on some benchmarks, and yet the same generative AI tools can flunk at simple counting exercises. Famously, many AI chatbots have historically failed to answer the question, 'How many *r*s in strawberry?' correctly.[82] While this particular example has been fixed, it is still easy to fool most chatbots. At time of writing, ChatGPT still thinks there are two 'r's in raspberry.

This apparent contradiction, where AI can be simultaneously clever and dumb, has been called Moravec's paradox. Moravec's paradox is very apparent in business applications. Even though AI can automate many complex business processes, some very simple things are still beyond reach.

Recently, my home insurer sent me an email asking me to update my home's details to include its elevation above ground (to assess flooding risk). My home sits directly on the earth and so has no elevation. But to tell my insurance company this, I had to go to their website, hunt down the relevant webpage, attempt unsuccessfully to engage their online chatbot, and eventually phone them up, waiting over 40 minutes to speak to someone. It should have been much simpler. This is an obvious case where I should have been able to quickly reply to the email and let AI do the rest. You don't need PhD-level maths skills for this task and yet AI couldn't do it.

So why are some things not yet automated by AI?

The answer is sometimes technical (large language models, for example, are not natural counting engines because of the way they work) but it is more often the case that an organisation isn't thinking about customer value in its AI use cases. Many businesses focus on automating processes and improving productivity, which may reduce costs but might not provide direct value to the customer. The classic example of this way of thinking is call centres that might be efficient from a financial point of view but customers hate them.

AI offers the potential for resetting this approach. As well as looking for productivity and cost reduction use cases, organisations should be taking a good look at customer value and how AI can support it. This requires a different way of thinking, approaching AI through a customer-focused lens. When it comes to AI, don't jump to solution mode too quickly. Any business leader can immediately see tasks that AI can automate, but these don't necessarily bring the best value to customers. To add customer value requires a deeper dive, starting with the problems that customers have, and then figuring out if AI can help.

WHAT IS CUSTOMER VALUE?

There are many ways to think about customer value. Eric Almquist and colleagues have identified 30 customer value propositions that capture what consumers want.[83] Examples include functional value (saves time, makes money, and provides information) and self-fulfilment value (gives a sense of belonging, offers hope, and provides motivation) (Fig. 7.1). According to this model, products and services can address four types of customer needs: functional,

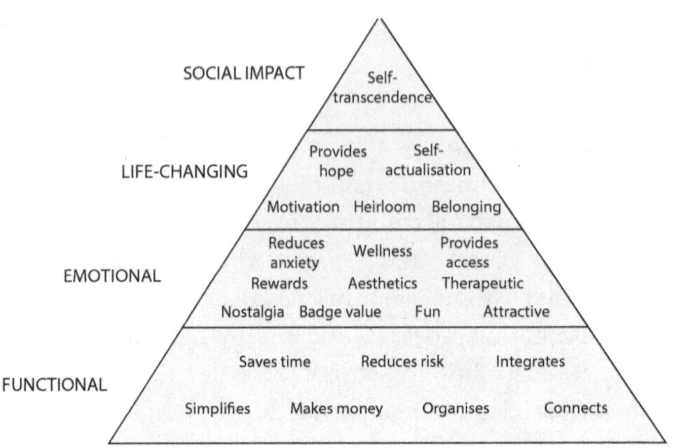

Fig. 7.1. Ways of thinking about customer value (adapted from Almquist *et al.*).[84]

emotional, life-changing and social impact. In general, the more elements that a product or service provides, the more likely a customer is to remain loyal to a business.

Note that Almquist's model of customer value goes well beyond straightforward efficiency for the customer. Going back to my home insurance example, there are clear functional values my insurer should provide me. A good insurance policy will *reduce my risk* and *save me money*. The insurance company should also *save me time* by providing efficient ways of communicating. All these are functional values, but a good insurer should also address higher-level values. For example, in times of stress, I want an insurance company that deals with me in an empathetic and considerate way, thus providing *therapeutic* value. My insurer might also *motivate* me to introduce preventive measures such as smoke alarms and security cameras. Even in the relatively simple case of home insurance, customer value can go way beyond productivity and cost.

Almquist's model is not the only way of thinking about customer value. More broadly, there have been decades of research

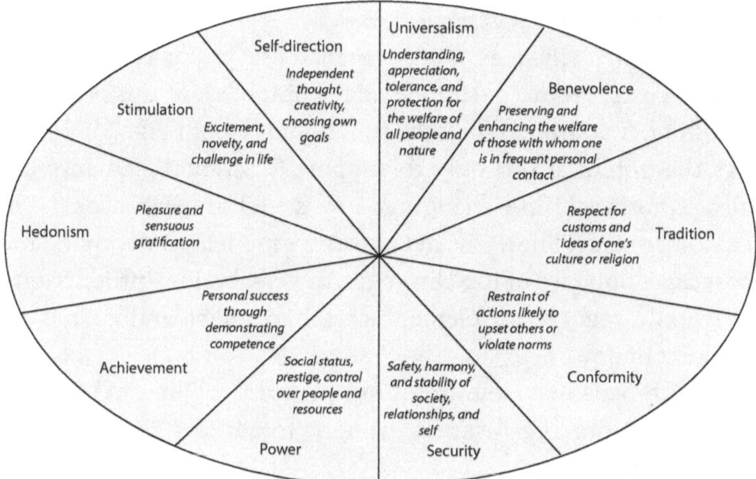

Fig. 7.2. Schwartz's theory of human values.

into *human values* and how they affect people's behaviour. The most well-known theory of human values is by social psychologist Shalom Schwartz.[84] Based on studies in over 60 countries, Schwartz identified 10 universal human values (Fig. 7.2). This model has applications in understanding consumer behaviour, marketing and branding and organisational culture.

Corporate values can also serve as a powerful lens for understanding and shaping customer value. They provide a guiding framework for what a company prioritises in its offerings, operations and relationships, directly influencing the kind of value customers perceive and experience. When corporate values are aligned with customers' own values and needs, they help ensure that products and services resonate more deeply, building trust, loyalty and long-term engagement. By viewing customer value through the lens of corporate values, organisations can deliver more meaningful, differentiated experiences that go beyond function to reflect shared purpose and principles.

AI AS VALUE ENHANCER

AI is often pitched as a way to increase productivity and/or reduce costs. From a customer value point of view, this is narrow thinking. It addresses only the functional layer in Fig. 7.1. But AI has the potential not only to support functional elements but also emotional, life-changing and social impact aspects of customer value. Similarly, in Schwartz's model, productivity and cost cover only two of 10 Schwartz's universal values: productivity is usually seen as an element of achievement and cost is an element of power.

Try broadening your thinking in terms of how AI can be applied in your organisation, using it to address a larger set of customer needs.

For example, a supermarket chain can use AI to measure sustainability across a supply chain more effectively, helping the organisation to develop a reputation as a sustainable business and to attract or retain customers. In this case, AI addresses Almquist's life-changing and social impact values, namely hope and self-transcendence. Customers who are interested in sustainability issues will be drawn to a supermarket with a strong track record on sustainability. Similarly, in Schwartz's model, sustainability is covered by the value of universalism.

As another example, a fitness company might offer its clients a new AI app to help them maintain motivation during a new diet or exercise regime (motivation in Fig. 7.1), thus addressing the customer value of achievement (Fig. 7.2). Such an app can be a competitive advantage in a crowded market.

Finally, an airline might create a partner program that offers loyal customers access to exclusive deals with luxury goods providers. This plays into the customer values of belonging and badge value in Almquist's model, and power in Schwartz's model, giving clients a way to obtain a higher social position.

SYSTEMATICALLY ADDRESSING CUSTOMER VALUE

When selecting AI use cases for your organisation, resist the urge to focus only on use cases that offer obvious productivity or cost benefits. By taking a broader lens, and thinking about customer value more widely, you may be able to identify AI use cases that offer better return on investment. To do this, it's a good idea to be systematic. I recommend the following three step approach:

1. Decide on a customer value list. You can use Almquist's or Schwartz's model or your own corporate values.
2. Tailor the model to your own context. The values in both Almquist's and Schwartz's models are quite abstract but will have more concrete meanings in your own business. For example, in aviation, universalism can be interpreted as sustainable fuels. In aged care, wellness might translate to absence of loneliness of residents.
3. For each value in the tailored model, brainstorm possible AI use cases that could positively impact the model. Like any brainstorming activity, this is best done with a diverse group of stakeholders and a good facilitator who can help participants think outside the box.

By following this systematic process you will be able to broaden your thinking on what AI use cases are best for your organisation. That's not to say that productivity and cost reduction use cases aren't important, but they are not the only game in town! Table 7.1 gives a few examples of AI use cases, tailored for customer values, across a range of industry sectors.

Table 7.1. Example AI use cases for adding customer value.

Value	Sector	AI use case
Universalism	Renewable Energy	Optimising energy storage and distribution through AI to enhance sustainability.
Benevolence	Mental Health & Well-being	AI-driven mental health chatbot providing empathetic support and mental wellness advice.
Tradition	Heritage Tourism	AI-based augmented reality recreating historical moments, preserving cultural heritage.
Security	Cybersecurity	AI for real-time threat detection and response to protect digital infrastructure.
Status	Luxury Fashion	AI for personalised luxury fashion recommendations, enhancing exclusivity and status.
Achievement	EdTech	Personalised language learning paths using AI to help users achieve learning goals.
Hedonism	Entertainment & Gaming	AI-enhanced real-time graphics rendering for immersive, aesthetically rich gaming experiences.
Simplicity	FinTech	AI-driven financial management tools simplifying budgeting and expense tracking.

QUESTIONS

- What are the top five customer values that drive your business?
- What AI use cases can you imagine that would support these values?

8

Can AI disrupt my industry?

WHEN I WENT TO WORK for NASA in 1999, Dan Goldin was its head. Goldin was famous for bringing a new mantra to this 40-year-old agency: *faster, better, cheaper* (FBC). At the time, the history of spaceflight was one of massive and complex missions, with a budget to match. Prior to Goldin's arrival, the agency's flagship program, Apollo, had required the coordination of 300,000 individuals working for 20,000 contractors and 200 academic institutions. At its peak, in 1966, Apollo's budget accounted for 4.4 per cent of the US federal budget.[85]

Spiralling costs and a lack of political will in Congress to sustain such levels of funding meant that by the time Goldin was appointed NASA Administrator in 1992, a new approach was required. Goldin coined this new approach faster, better, cheaper.

As a 26-year-old joining NASA – my first job after a PhD – I heard the FBC mantra everywhere. It was proclaimed in the corridors and referenced in meetings as the answer to all of NASA's challenges. It was the perfect example of organisational vision: pithy, easy to understand and, as a result, quickly embedded in the minds of every NASA employee, even a rookie computer scientist like me.

FBC was Goldin's way of capturing how NASA needed to change. It reflected a shift from large, expensive, single missions to smaller, more frequent missions with tighter budgets. In many ways, FBC was successful. It reduced average mission development times from six to eight years (typical for Apollo) to about three to four years. Costs per mission were cut by as much as 50 to 75 per cent, without sacrificing scientific value.

By doing things faster, better and cheaper, NASA disrupted itself. But the approach was not without its failures – missions like Mars Climate Orbiter (1999) and Mars Polar Lander (1999) were lost due to errors attributed to the push for speed and efficiency. Still, FBC changed the agency's culture and prepared it for the next era of space exploration.

The true disruption, however, came later. By the late 2000s, faced with continued high costs and aging infrastructure, NASA fundamentally changed its business model. It moved away from missions entirely dominated by NASA to a public–private partnership model. This shift saw the rise of companies like SpaceX, which would, in 2012, become the first private company to send a cargo spacecraft to the International Space Station.

SpaceX was truly disruptive. It fundamentally changed the economics and technology of space exploration, breaking NASA's monopoly on spaceflight and forcing the entire aerospace industry to rethink how it designs, builds and launches spacecraft. Its innovations introduced unprecedented cost reductions, reusability and rapid iteration, reshaping the space industry into a competitive, fast-paced sector dominated by private companies rather than governments.

SpaceX's real innovation was that it approached space exploration *differently*. It revolutionised the concept of rocket reusability. Traditional rockets were single use, discarded after launching their payload. SpaceX changed that. It challenged the traditional aerospace model by vertically integrating most of its operations and adopting an iterative, Silicon Valley-style development approach. In short, SpaceX didn't just make the industry faster, better and cheaper, its innovations overturned decades of established thinking and changed the very way that the industry worked.

FASTER, BETTER, CHEAPER AI

AI can undoubtedly make your organisation faster, better and cheaper (if done right). The first chapter in this section of the book showed how AI can lead to productivity benefits, making your organisation *faster*. The second chapter discussed how AI can lead to cost reductions and potentially revenue increases (i.e. *cheaper*). And the third chapter looked at how AI can provide *better* value to the customer.

Faster, better, cheaper is great. If that's all you get from AI, you're doing well. But the true value from AI comes when it allows you to be truly disruptive, to do things that simply wouldn't be possible otherwise.

Achieving true disruption with AI is far from easy (and not always necessary), but it's worth giving some thought to whether a more fundamental disruption is possible, because it's a helpful way of understanding the benefits that AI can really bring to your business. Too often I see companies adopt AI in a very narrow sense. They quickly fixate on existing tasks that can be automated. But applying AI only to speed up an administrative task is a lost opportunity. AI can potentially turn your business model on its head, enable you to enter markets that are otherwise out of reach, and force you to revisit your thinking about how your organisation works.

To take a simple example: in 2025, three companies – openAI, Google and Perplexity – all released their versions of 'deep research', an AI tool that can generate research reports, citing relevant literature, on any topic of your choosing. With a single prompt, such as 'create a structured report on the economic benefits to different European countries of blockchain technologies', the AI can generate, almost instantaneously, a 12,000-word essay that neatly summarises all the research on this topic. The result is magical and has been compared to PhD-level intelligence. If report writing is part of your job, AI is certainly faster and cheaper. It is arguably better too.

But it isn't disruptive. A disruptive thinker asks why generate a 12,000-word essay in the first place?

The corporate world is replete with lengthy reports, yes. But simply getting AI to write reports that no one will read is hardly disruptive. The disruptive thinker asks if there is a different way. Maybe 12,000-word reports aren't the best way to communicate complex subjects. Maybe AI could enable completely different ways of communicating information beyond the written word. Maybe you don't need those reports at all and instead should restructure your business processes to eliminate them. Arguably, the value of AI isn't the technology itself but the fact that it forces you to ask questions about the value of your business processes. That is true disruption.

EXAMPLES OF AI DISRUPTION

There are examples where AI has enabled *different* business models, rather than simply making existing business models more efficient.

TikTok is a case in point. TikTok is a Chinese-owned social media app launched in 2017. It quickly became one of the most widely used social media applications, popular with young people. It is now in the top five most used social media platforms, with 1.5 billion active monthly users.[86] Whilst TikTok is known for its short-form video content, the disruptive model of TikTok was not its content but the algorithms behind its feed. Traditionally, social media platforms served up content based primarily on followers.

The assumption was that if I follow someone, I will like the same things as they like. Most social media applications – Facebook, X, LinkedIn – still work this way. You follow someone and then get to hear what they are saying.

TikTok works in a fundamentally different way. Its content is served up based on user behaviour, not followers. An AI-powered recommendation algorithm learns user preferences based on interactions with the platform – likes, comments, shares, and how long a user watches each video. The result is that content is delivered according to what the user likes rather than who the user follows. TikTok wasn't the first platform to use AI to deliver video content but it took things to a new level, with better personalisation for short content and faster learning models. Delivering content based on user behaviour wasn't just faster, better and cheaper; it was a different way of doing things.

Another example of AI enabling different ways of doing things is online dating. The first successful commercial online dating platforms appeared in the 1990s with kiss.com and match.com. Dating online was, of course, disruptive in itself. It allowed you to meet a partner without ever needing to leave home. Online dating was *different*. Early online dating platforms relied on questionnaires and personality profiles to assess user preferences and create matches. It wasn't until the 2010s that online dating underwent another disruption when large-scale neural networks began to be used to suggest matches based on users' behaviour on the platform – who they interacted with, profile views, message responses, and swipe behaviour. Nowadays, almost all online dating platforms use AI to recommend a potential partner.

AI led to new business models in online dating, with a move away from long-term subscription models to shorter-term transactions with advanced AI features offered as premium services. The driver for introducing AI into online dating wasn't necessarily to be faster, better or cheaper. The real impact was to change the business model. AI wasn't used to match personality tests more efficiently; it eliminated the personality tests entirely.

In a completely different field, another AI-driven disruption is AlphaFold.[87] AlphaFold is an AI system that can accurately predict the structures of proteins based on their amino acid sequences. Predicting protein structures had been one of biology's greatest unsolved challenges for 50 years. Solving it, as AlphaFold has done, enabled radical advances in drug discovery, personalised medicine and disease research. Traditional methods could take months to predict a single protein, making progress towards understanding the millions of protein structures that exist very slow. AlphaFold reduced the time to predict a protein structure to a matter of hours and, in some cases, worked where traditional methods were impossible.

One of the most striking examples where traditional methods failed is with the nuclear pore complex (NPC) – a massive protein assembly that regulates the transport of molecules between the cell nucleus and cytoplasm. For decades, scientists struggled to determine its full three-dimensional structure. With AlphaFold, researchers were able to predict high-resolution structures of previously unresolved NPC proteins. This breakthrough has provided crucial insights into how the nuclear pore functions, which is essential for understanding its role in diseases like cancer and neurodegenerative disorders.

AlphaFold wasn't just faster, better and cheaper – it changed the rules of the game.

AI AS A DISRUPTER IN YOUR BUSINESS

Being disruptive is hard, and it's not always necessary. Sometimes it is sufficient to be faster, better and cheaper. But if you are interested in being disruptive, how do you go about it? It helps to have a structured approach.

There are many different frameworks you can use to stimulate disruptive thinking. Examples include 10X Thinking, the Blue Ocean Strategy, Design Thinking, the Innovation Ambition Matrix and Lean Startup.[88] Whichever framework you use, there are a few principles to bear in mind when trying to come up with truly disruptive solutions:

1. *Start with organisational strategy.* You are aiming to disrupt within the boundaries of your organisational strategy, not disrupt the strategy itself. A car company can disrupt by inventing self-driving cars, but it shouldn't disrupt by inventing a new type of bicycle. Disruptive thinking needs constraints, and your strategic goals usefully bound your thinking; otherwise, there are too many possibilities to consider. When it comes to disrupting with AI, look for strategic goals where AI can enable something new and different to happen or can solve a problem that's previously been unsolvable.
2. *Bring in outsiders who think differently.* If you've been working in the same organisation, or even the same industry, for a long time, it's very hard to imagine how things could be different. You need people who are far removed from the usual conversations. This could be someone external to your organisation, but it could also be 'outsiders from within'; that is, people who are radical thinkers, new to the organisation, or are willing to challenge the status quo.
3. *Look for creative thinkers.* It goes without saying that you want creative thinkers when it comes to disruption. You'll need other personality types to implement the solutions, but creatives are the most important when it comes to breaking boundaries and having the initial ideas. Look for creatives within your own organisation who understand the business

but will approach things laterally. If useful, you can use thinking styles assessment tools like HBDI[89] to identify which of your employees tend to think more creatively.

4. *Empower people to challenge assumptions.* In many organisations, staff are reluctant to ask why. As a result, long-held assumptions go unchallenged, even if the original reason for them has long since vanished. In a disruptive thinking process you need to empower people to challenge these assumptions and ask, 'why do we do it this way?' There are some neat tactics for encouraging this – for example, at each meeting, allocate one person to be the 'naysayer' and empower them to ask the questions that no one else wants to.
5. *Include the right kind of AI expertise.* This might be the trickiest part to get right. You need someone in the room with an AI background. If you don't, you are likely to come up with fantastical and unfeasible AI solutions. Someone needs to bring you back to reality. On the other hand, you don't want to be too constrained. Many technical experts are conservative and not very creative. You need to find someone within your organisation who has the technical background and is naturally creative.

CONCLUDING THIS SECTION

This section of the book has covered four aspects to think about when applying AI to your organisation: productivity, cost, customer value and disruption. These can be used to frame conversations about the best use cases for your business. Don't limit yourself to just one of these aspects. It's easy to fixate on just productivity or cost, but it is often customer value or disruption where the true benefits of AI become apparent.

In the next section of the book, we'll delve deeper and understand how to develop an AI strategy, how to structure your organisation around this strategy and how to make sure AI is working for you.

QUESTIONS

- Can you think of any truly disruptive innovations in your industry that have achieved lasting impact?
- What's the one problem of your industry that if solved, would truly change it? Could AI help to solve this problem?

Top tip: Disruptive thinking

There is an art to being a disruptive thinker. But it also helps to have a structured framework. I'm a big fan of design thinking as an approach to ideation, which focuses on understanding user needs, behaviours and motivations. When it comes to disruptive AI innovation, focusing simply on users is not sufficient. Instead, the focus should be on identifying previously unsolvable problems that AI can solve. Think of the apocryphal quotation about the idea for automobiles, attributed to Henry Ford: 'If I had asked people what they wanted, they would have said faster horses.'

Figure 8.1 is a five-step process you can use to encourage disruptive AI innovation. The framework expands on design thinking but shifts the focus from user-driven incremental improvements to radical, AI-enabled breakthroughs that weren't possible before.

Explore. Ask: What fundamental challenges in our industry have remained unsolved? Identify industry limitations: what problems are so complex, time-consuming or data-intensive that no human solution has worked? Assess emerging AI capabilities: are there recent AI advances that could unlock new ways of solving this problem?

Reframe. Identify assumptions in your industry that AI might invalidate. Challenge status quo thinking: what if AI could replace entire workflows, industries or business models?

Invent. Map AI's unique strengths: pattern recognition, autonomous learning, generative capabilities, real-time adaptation, and predictive intelligence. Ask: what can AI do here that no human ever could? Brainstorm AI-enabled solutions that would fundamentally change how an industry operates.

Prototype. Create AI-driven prototypes to test concept viability, not just usability. Measure for industry transformation, not just customer satisfaction. Consider second-order effects: does this invention create new behaviours, markets or entire industries?

Scale. Plan for mass adoption: how will this AI system integrate into or replace existing processes? Anticipate industry resistance: what regulatory, ethical or business model challenges could arise? Scale up with ecosystem thinking: how can this AI solution expand into adjacent markets?

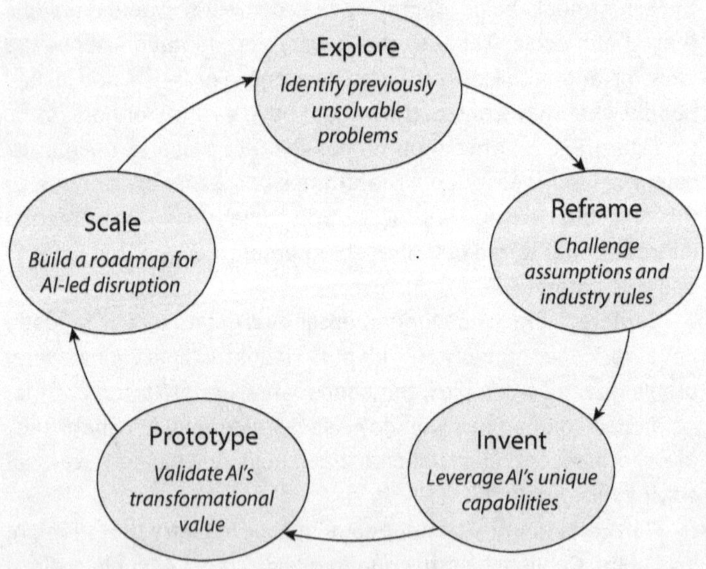

Fig. 8.1. Process for disruptive AI-led innovation.

Part III
AI strategy

Part III of this book moves us from the 'why' of AI to the 'how'. By now, you should have a good understanding of why your organisation wants to adopt AI – whether it's to increase productivity, to reduce costs, add customer value or even disrupt your industry. The natural next question is how you are going to achieve this. I will cover how to come up with a good strategy and how to set up organisational structures to ensure your strategy is a success.

There are three chapters in this section of the book:

- ***AI strategy.*** **What does a good AI strategy look like and how will you develop one? What strategic choices must you make when considering AI** adoption and how will you make trade-offs that set you up for success?

- *Organisational structure.* Who is responsible for successful AI adoption in your organisation? Where does that person sit in the organisational structure and how will you structure a team around him/her? What are the different models for structure when leading an AI transformation?
- *AI governance.* Setting up effective oversight is crucial to the success of AI. How will you ensure you are managing risks appropriately? Are you aware of your compliance responsibilities? And how will you keep on top of the rapidly changing AI regulation landscape?

9

What should my AI strategy be?

A CIO FRIEND OF MINE once called me up, all flustered. He worked for a large education services company. He had a background in tech but didn't know much about AI. His company was very traditional and wasn't used to introducing cutting-edge technologies. Until now.

'The board are serious about AI,' my friend blurted out. 'They want me to come up with an AI strategy. What should I do?'

This is a question I am often asked. 'What should my AI strategy be?' Usually quickly followed by 'and how do I go about developing it?'

It can be a challenge for organisations – especially non-technology ones – to develop an AI strategy. Organisational strategy isn't easy at the best of times. Before embarking on developing a strategy, therefore, it's important to be clear what a strategy represents.

University of Toronto Professor Roger Martin says: 'Strategy is an integrative set of choices that positions you on a playing field of choice in a way that you can win.' This is the essence of strategy. It is a set of *choices* about possible future worlds. The hard part of strategy is that those alternative futures come with a great deal of uncertainty. And, when it comes to AI, there's even more uncertainty than usual. The technology is changing so fast that it's impossible to know what AI will look like even in a few years' time. As a result, it's hard to predict how your industry will evolve. Will it resemble the current state but be faster and more productive? Or will the industry have been disrupted by AI in ways you can't imagine?

An AI strategy then is a set of choices that determine how your organisation will position itself and how AI can help with that positioning.

And the very first choice you need to make is ...

CHOICE NUMBER ONE – YOUR AI POSTURE

The first decision point is your *AI strategic posture*. A strategic posture is a high-level agreement on your market positioning when it comes to AI. Do you see yourself as an early adopter of AI, leveraging it to disrupt your industry, or are you content to sit back and wait, adopting AI technologies only once they are stable and have been proven in your industry? Hugh Courtney and colleagues have defined three strategic postures, designed to help companies navigate through times of uncertainty.[90] I have adapted these postures to the context of an AI strategy:

- *AI disruptors* use AI to shape how an industry will evolve. For example, when Netflix transitioned from mail-out discs to video streaming, it pivoted its movie recommendation systems, moving away from explicit user ratings to AI-powered suggestions based on viewing history, watch time and clicks. It set the gold standard for AI-driven content recommendations, which rival streaming services later adopted. Netflix used AI to shape the video streaming industry, but it also shaped e-commerce and retail, showing how AI recommendations could be used to sell products. AI disruptors set the standards of how AI will change an industry, creating new market demand and reshaping the industry.
- *AI adaptors* work within current industry structures and try to optimise their position within it using AI. They are not looking

to reshape an industry but to maximise chance of success within current norms. For example, a telecommunications company might use AI to improve customer service response times, giving it a market edge, but this doesn't change the services that telecoms providers offer. Whereas AI disruptors seek out ways that AI can do things differently, AI adaptors are keen to adopt AI but are content to focus on making the existing system more efficient.

- *AI watchers* commit only to incremental investments, waiting to see what others do or how the AI landscape will develop, getting on board just enough to satisfy their fear of missing out. For example, a research organisation might support staff to use generative AI tools in their everyday work but be reluctant to invest in more advanced AI to disrupt the scientific process. Whereas AI adaptors are active investors, AI watchers are more conservative. It's not that they don't invest in AI at all but they prefer to avoid big bets until the environment has stabilised.

The first choice in your AI strategy should be whether you want to be an AI disruptor, adaptor or watcher. This is a critical strategic decision as it will affect every other aspect of your strategy. If you decide to disrupt, you may have to fundamentally revisit what your organisation is trying to achieve. The risk is high, but so is the payoff. If you are a watcher, it's sufficient to assess your current business goals to see if there are some quick AI wins. The risk here is that you'll get left behind by your competition.

To be clear, one choice is not better than another. I'm not encouraging all organisations to be disruptors. In many cases, watch and wait is a perfectly appropriate course of action. But you ought to make a considered and definite choice.

It is worth putting a good deal of thought into your strategic posture and make sure all key stakeholders agree. Without this, there is a risk that your AI strategy will be constantly derailed because participants don't agree on the strategic posture.

CHOICE NUMBER TWO – WHAT PROBLEM IS AI SUPPOSED TO SOLVE?

AI is meant to solve a problem. Let me repeat that: *AI is meant to solve a problem.* That might sound obvious, but you'd be surprised how many organisations I've worked with that haven't thought clearly about the problem that needs solving. Or they haven't thought about *which* problem needs solving, jumping straight to solution mode for the first problem they can bring to mind.

As a cool technology, there's a tendency just to adopt AI *somewhere*.

Back to my CIO friend. 'What kind of AI strategy do the board want from you?' I asked him, hoping to tease out the board's strategic posture. 'I don't know,' replied my friend. 'They just want the executive team to do something with AI.'

'I see.' I paused. 'Perhaps you should talk to them?'

When he talked to the board, they agreed that disruption wasn't realistic. They were AI adaptors.

'Now what?' my friend asked.

'Well,' I said, 'do you know what problem you are trying to solve?'

There's no point randomly looking for use cases where you can apply AI. If you do that, you're not being true to your strategic posture. You'll end up focusing on quick wins – which is fine if you are an AI watcher but not otherwise.

Before selecting AI use cases, it's important to be clear on the problem. The size of problem you focus on depends on your strategic posture. For an AI disruptor, the problem may be your entire business strategy. For an AI adaptor, the problem will be captured in one of the key goals in your business strategy. You won't want to change the fundamentals of the strategy; rather, you'll use AI to help deliver on a goal. For an AI watcher, the problem is probably too small to appear in your business strategy at all. In all three cases, problem clarity is key.

In the rest of this section, let's assume you are an AI adaptor. As a disruptor, you need to undertake a fundamental rethink of how you do business, but as an adaptor, the best place to start is with your existing strategic goals.

I recommend a simple process to enumerate the problems your organisation has and assess whether there's an AI use case that can help solve the problem. Table 9.1 lists examples of strategic business goals, the underlying problems they are intended to address and potential AI solutions that could help solve them. Strategy documents often don't explicitly capture problems, but they are there. Every strategic goal is intended to solve a problem: you just need to reframe your goals to make it explicit.

Table 9.1. Mapping out how AI can help to solve strategic problems.

Example strategic goal	Problem it is trying to address	Potential AI use case solution	Will the use case solve the problem?	Is the use case feasible? Can we deliver it? What's the return on investment?
Be employer of choice	Losing talent due to delays in recruitment process.	AI recruitment system to speed up processing and improve candidate care.	Maybe. Need to understand current bottlenecks to ensure AI is directed at the right place.	Don't know. Staff are resistant to changing systems due to poor change management in the past.
Reduce customer response times to 24 hours	Too few call centre staff leading to massive backlogs, particularly for routine queries.	Chatbots to handle routine queries.	Probably. Need to test to what extent AI can answer routine queries.	Yes. CIO has experience implementing such systems from previous role.
Increase staff productivity by 10%	Too much bureaucracy causing staff frustration and burnout.	Generative AI to automate preparation of documents.	Not really. AI can provide speed-up but more effective to re-engineer business processes.	

continued

Table 9.1. continued

Example strategic goal	Problem it is trying to address	Potential AI use case solution	Will the use case solve the problem?	Is the use case feasible? Can we deliver it? What's the return on investment?
Double sales within 12 months	Customers not returning for future sales.	AI marketing solution to engage existing customers and recommend products.	Maybe. AI is good at recommending products but needs high-quality data to make good recommendations. May require work first to collect and prepare historic sales data.	Maybe. Will need to partner with a consultant to address the data quality issue.

By following a structured process to evaluate AI as a solution against your strategic goals, you'll end up with a list of potential AI use cases strongly aligned to your business strategy. The next step is to prioritise these use cases.

CHOICE NUMBER THREE – PRIORITISE

I spoke a few more times with my CIO friend. By this point, he had worked with his team to come up with a list of possible AI use cases, all neatly aligned to strategic goals.

'That's fantastic,' I congratulated him.

'Is it?' he asked, despondently. 'There are so many possible uses. I just don't know where to start. We can't do them all.'

'Indeed, you can't,' I replied. 'Looks like it's time to make some choices.'

Once you have identified the problems that AI *could* solve, you need to prioritise them. This requires a thorough assessment of six dimensions (Fig. 9.1). By evaluating AI use cases along these six dimensions, you'll be able to make an informed choice which use cases to pursue.

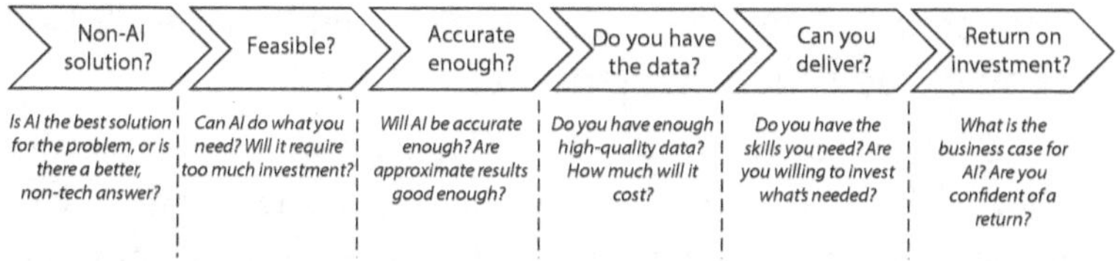

Fig. 9.1. Six questions to ask when evaluating AI use cases.

Dimension 1: Right solution for the job. Is AI the best solution for the problem or is there a better, non-tech answer?

There's so much hype around AI that every problem can start to look like a nail, where the AI is a hammer. But not every problem is a nail. And just hammering at something with AI isn't going to improve things.

A critical choice in the next step of your AI strategy is to step back and ask if AI is really the solution. Sometimes, the problem isn't one that has an easy technological fix. For example, are your business processes overly bureaucratic, requiring staff to spend hours each week creating paperwork and seeking approvals from multiple managers? Is this a problem AI could solve? Possibly. You could automate the generation of that paperwork. But where would that get you? You would still have inefficient processes. Wouldn't it be better to fix the processes themselves? In this case, I'd argue that AI isn't the right way to fix the problem. It's much better to re-engineer the way your organisation works.

It is perfectly OK to decide that AI is not the answer. That's a good thing, because it narrows down the AI use cases and helps you decide where AI can genuinely help. It is far better to eliminate AI as a solution early, rather than pushing ahead only to discover problems later. Don't be tempted to use AI as a band-aid to paper over the cracks in your business. Instead, prioritise cases where AI can genuinely and honestly offer value.

Dimension 2: Feasibility. Can AI do what you need? Will it require too much investment?

Not all AI use cases are feasible. It is easy in a brainstorming session with strategy execs to come up with a magical AI use case that will solve all your problems. But remember: AI is maths, not magic. Can it really solve your problem?

Even where AI is a *potential* solution, you need to ask if that solution is feasible within your organisation. This dimension looks at feasibility in general terms. At this point, resist the urge to dive deep into your organisation's capability to deliver; that comes later. For now, simply ask if the AI solution is even possible in an organisation like yours.

For example, if you've identified the need for an AI to recommend new products to customers based on past purchases, this is a well-understood AI use case that is relatively easy to implement. It should be feasible. However, if you're a pharmaceutical company wanting to use AI to develop a new cancer drug, that's a major R&D effort that could take years, if not decades, to achieve. It isn't always easy to determine if a use case is feasible. Suppose you are a logistics company that wants AI to optimise your transportation costs. Is that doable? Do you know what existing solutions are out there? If you need to build your own solution, do you know how much it will cost? At the end of the day, if your AI idea isn't feasible, choose another solution.

Dimension 3: Accuracy. Will AI be accurate enough for your needs? Do you need perfection or are approximate results good enough?

Modern AI systems (i.e. data-driven AI) are never 100 per cent accurate. That's a fact. Remember that AI systems are large-scale statistical pattern-matching systems. And statistics is probabilistic. AI will never get things right all the time.

That might initially sound like bad news. But for many applications, you don't need 100 per cent accuracy. An AI that recommends movies on a streaming service can afford to make a bad recommendation occasionally. It might not be great customer experience but no one dies from watching a bad movie (unless, of course, you're a character in *The Ring*). But for many applications, perfect – or near perfect – accuracy is essential. If you are a GP, you might have a great idea to triage routine patient concerns using a chatbot, but that could get you into hot water quickly if you misclassify a patient who then gets seriously ill.

A key question to ask for the problem you are trying to solve, therefore, is what error levels can be tolerated. Bottom line: if you need very high levels of accuracy, think carefully before using AI.

Dimension 4: Data availability. Do you have enough high-quality data to train your AI? How much will it cost to get your data into shape?

AI systems are data hungry. Without large amounts of reliable data, your AI won't be able to do much. You can have the best AI idea in the world but if you don't have any data, it will stay just that: an idea. So when assessing AI use cases, ask if you have a suitable dataset. Is your dataset large enough? Is the quality of your data high enough? The biggest barrier to using AI for many organisations is the investment needed to clean up data to ensure the information doesn't contain biases and has enough good-quality information so the AI can do its thing. Can you access the data? You might have a great customer dataset but what agreements are in place with your customers to let you use that data, and for what purpose? Even if you technically can use the data to train AI, do you want to? Should you inform your customers? What if they don't want you to use the data? Maybe you can't access the data because of privacy laws or sector-wide conventions.

There is a lot to think about when it comes to data.

Dimension 5: Delivery. Do you have the skills you need to implement the AI use case or can you acquire them? Is your organisation willing to invest what's needed?

By this point, you've identified a problem and you're confident there's an AI solution. But can you deliver it? Do you have the resources and skills within your team to develop, test and deploy the AI? Are the people you need available?

It is important to understand that AI is not an off-the-shelf technology. It takes significant effort to deploy a real-world AI system. It is certainly not as simple as paying a small subscription to a generative AI vendor and then just using it.

You will need plenty of AI experience – this can come from within your organisation if you have it; otherwise, you'll need to partner with another organisation.

Dimension 6: Return on investment. What is the business case for your AI solution? What investment will you need to make and are you confident of a return?

This dimension is where it gets real. You need to spend money to make money. But exactly how much will you need to spend? And are you sure – *really* sure – that you'll recoup your investment?

Digital transformation projects are notorious for cost overruns, and AI projects are no different. It can seem that AI is magic tech that is cheap and will supercharge your business, but that's almost never the case. The levels of investment for an AI use case can be significant. These projects are also hard to estimate. AI is an emerging technology. There may be no one in your organisation that has done this kind of project before so you won't have historical figures to rely on. And you can't necessarily believe what tech vendors will tell you.

Unfortunately, there is no easy way around this problem. My only advice is to spend a good deal of time mapping out the business case for AI properly. Critique the case from all angles. Consider multiple scenarios. What is the gain if it works? What have you lost if it fails? Be a sceptic. Avoid being enamoured by the technology hype. Ask yourself hard questions. It is better to know the real costs now than to discover them later.

CHOICE NUMBER FOUR – HOW DO YOU EXECUTE AN AI STRATEGY?

Congratulations. You've got to the point of execution! You've done a lot of hard work in agreeing a strategic posture and in understanding which AI use cases can provide most value to your organisation. You have a prioritised list of feasible and realistic use cases that you can deliver.

All that's left to do is implement them. Easy, right?

Your fourth (and final) choice is actually a set of choices. There are many ways you can go about executing an AI strategy. There is no single *right* way. How you fare will depend on your organisation's culture, style and appetite for change. How you go about implementing AI use cases is a core part of your AI strategy. It will determine what kind of governance structure you need, how you will monitor success and where you will invest time and resources.

Table 9.2 outlines tactics for executing an AI strategy and describes the pros and cons of each. I elaborate on these tactics below. For executing your strategy, think about which of these tactics you will use. You can use a combination if you wish.

Table 9.2. Tactics for executing an AI strategy.

Trade-off	Tactic	Pros	Cons	Example
Centralised v decentralised	**AI centralists**	- Ensures alignment with organisational goals. - Standardised processes and governance. - Easier to scale up successful initiatives. - Facilitates centralised resource allocation.	- Can stifle innovation and creativity. - May be slower to respond to localised needs. - Higher dependency on central teams.	**Retail company**: Implements AI for system-wide inventory management, aligning all stores to a central strategy.
	AI decentralists	- Encourages innovation and creativity. - Empowers employees to experiment with AI. - Adaptable to localised needs. - Can lead to grassroots innovation.	- Risk of inconsistent practices. - Lack of standardisation may create inefficiencies. - Potential duplication of effort.	**Retail company**: Makes generative AI tools available for employees to reduce admin work and innovate individually.
Level of managerial oversight	**AI bureaucrats**	- Ensures alignment with organisational goals. - Provides clear accountability and metrics. - Reduces the risk of misaligned initiatives.	- May stifle creativity and innovation. - Lengthy bureaucratic processes can delay project execution. - Risk of disengagement from creative teams.	**Large corporation**: AI projects must define ROI, report monthly to the executive team, and align strictly with corporate goals.

Trade-off	Tactic	Pros	Cons	Example
	AI skunks	- Encourages agility and innovation. - Allows for rapid experimentation and development. - Minimises bureaucratic delays. - Can foster ground-breaking ideas.	- Risk of project overruns or misalignment with organisational goals. - Lack of accountability may lead to inefficiencies. - May face challenges scaling up successful ideas.	**Tech company**: A skunkworks team develops an AI-powered prototype outside normal corporate constraints to explore untested market opportunities.
Level of criticality	**Mission-critical pilots**	- Demonstrates organisational commitment to AI. - High impact on competitive positioning. - Generates significant value if successful. - Encourages robust planning and execution.	- High risk if the project fails. - Requires substantial investment. - Potential for reputational damage if outcomes are negative.	**Healthcare company**: Uses AI to diagnose critical illnesses, ensuring precision and reliability in life-saving applications.
	Test pilots	- Low-risk and low-cost entry point to AI. - Ideal for learning and experimentation. - Provides insights without major disruptions. - Builds organisational confidence in AI.	- Limited immediate impact. - May not fully demonstrate AI's potential. - Results might not justify scaling.	**Healthcare company**: Tests AI for scheduling patient appointments to improve operational efficiency without high stakes.

continued

Table 9.2. continued

Trade-off	Tactic	Pros	Cons	Example
Quick wins or focus on value	**AI tweakers**	- Quick implementation with minimal disruption. - Low cost and effort. - Focused on solving immediate problems. - Easier to demonstrate short-term ROI.	- Limited scope may not fully utilise AI's potential. - Incremental impact on competitiveness. - May require frequent updates or expansions.	**E-commerce company**: Introduces an AI chatbot to handle FAQs and basic customer queries.
	AI overhaulers	- Enables systemic, organisation-wide transformation. - Long-term competitive advantage. - Fully leverages AI's potential to optimise operations. - Aligns AI with overall business goals.	- High initial cost and resource demand. - Longer implementation time lines. - Significant cultural and operational disruption.	**Manufacturing company**: Integrates AI into its supply chain, from demand forecasting to inventory management.
Problem or technology focused	**AI problem solvers**	- Ensures AI aligns directly with business needs. - Increases likelihood of measurable ROI. - Reduces risk of wasted resources. - Focused on practical, targeted solutions.	- May limit innovation to 'known unknowns'. - Risks missing opportunities beyond immediate problems. - Slower to adopt emerging technologies.	**Logistics company**: Implements AI to optimise delivery routes, addressing inefficiencies identified as a pain point.

Trade-off	Tactic	Pros	Cons	Example
	AI technologists	- Encourages experimentation and innovation. - Can unlock 'unknown unknowns' and unexpected opportunities. - Builds a forward-looking, tech-savvy culture. - Positions the company as a leader in cutting-edge tech.	- Risk of focusing on technology for its own sake. - Potential misalignment with business needs. - Higher likelihood of wasted investment.	**Tech-savvy retailer**: Adopts generative AI to experiment with personalised product recommendations without a clear ROI target upfront.

Centralised versus decentralised

An *AI centralist* implements AI strategy in a centrally controlled, top-down way. This involves a tightly defined AI strategy, a central AI leadership team, clear governance mechanisms and rigorous change management to drive transformation across the organisation. *AI decentralists*, on the other hand, adopt a more experimental approach, making AI tools available to employees and encouraging them to innovate on their own tasks in a bottom-up manner. There can still be central coordination, but the emphasis is on giving employees freedom to try things out. The decentralised approach can be attractive for AI because, as an emerging technology, it is unknown *a priori* how AI can best be applied. AI decentralists lean on the wisdom of the crowd to innovate with the tools available and to find solutions that a centralised leadership team would not.

Level of managerial oversight

Levels of oversight for an AI strategy can vary. At one extreme, there is close monitoring, with tightly defined KPIs and regular reporting to the executive team and/or board. This can lead to a bureaucratic process that risks stifling the innovation needed for success in AI. I therefore call this approach *AI bureaucrats.* At the other extreme, *AI skunks* are given free reign and lack of bureaucratic oversight in a skunkworks project; that is, a small, autonomous team given permission to operate outside the usual corporate constraints. Skunkworks projects can lead to increased agility and innovation but they risk project overruns if there is insufficient oversight.

There is overlap between the level of oversight choice and centralisation versus decentralisation. A centralised approach is more likely to have bureaucratic oversight. This doesn't have to be the case, however. You can have a centrally defined strategy but allow employees to innovate within that strategy in a skunkworks-like way.

Level of criticality

When executing your AI strategy, how seriously will you treat its success? Will you go 'all in' so that business success depends on successful execution of your AI strategy, or will you tread more cautiously and focus only on low-risk pilots?

A common approach is for companies to start with AI pilot projects as a way to learn about the technology and improve their understanding of return on investment. However, there are different types of pilot projects in which a company can invest. *Mission-critical pilots* are high-stakes pilot projects where failure is not an option as the outcomes significantly impact the organisation. A healthcare company using AI to diagnose critical illnesses would be deemed mission-critical because it could make or break the company. *Test pilots* are lower-stakes projects focused more on learning and exploration. An example would be the same healthcare company using AI to schedule patient appointments. Whether you choose to start with mission-critical or test pilot projects will depend on your organisation's risk appetite and technology maturity.

Quick wins or focus on value

When executing on AI, companies have different appetites for the breadth of change they are seeking. *AI tweakers* look for quick wins, where they can apply AI to address specific, immediate problems with minimal disruption to existing processes. A company might, for example, introduce an AI chatbot to field answers to frequently asked questions. This can be a relatively quick and low-cost way of adopting AI, but it might not make full use of the opportunity. *AI overhaulers*, on the other hand, look at the business holistically and introduce organisation-wide change to achieve systemic benefits. For example, a manufacturing company might integrate AI across its entire supply chain. There is a correlation between strategic posture and the focus on quick wins or longer-term value. They don't need to align, however. Even a disruptive company might initially go for a few quick wins to gain stakeholder buy-in.

Problem or technology focused

I have emphasised throughout this chapter the importance of first understanding the problem you are trying to solve. There is an argument to be made, however, that it sometimes makes sense to start with the technology and worry about the problem later. Generative AI, for example, is a very new technology and organisations don't understand how best to use it. As a learning exercise, it can be useful to make generative AI tools available to staff and see what they do with it. I call proponents of this approach *AI technologists* because it is usually technology enthusiasts who follow this style of implementation. Their excitement about AI can be contagious and can even inspire a whole organisation to start experimenting. *AI problem solvers*, on the other hand, focus squarely on the problem to be solved. Both approaches can succeed, although the former needs to be led by a charismatic technology superstar who can galvanise a community. Without this, you might waste a lot of time hoping your staff will adopt AI but they never do.

EXECUTING AI STRATEGY IS HARD

Let's revisit my CIO friend to see how he's getting on. I hadn't heard from him for a few weeks, but then he called me up one day, all excited.

'We've done it!' he exclaimed. 'I presented the AI strategy to the board and they've endorsed it.'

'That's great,' I said. 'Have you decided how you'll execute it?'

There was a long pause on the other end of the line. Eventually, my friend stuttered, 'Execute it? No ... I haven't really thought about that yet. You're not going to tell me there's more than one way to do that, are you?'

'I'm afraid so,' I replied, gravely. There are always more strategic choices to make.

QUESTIONS

- Think about your own organisation. Which AI strategic posture do you think would fit the culture best?
- Come up with a list of two or three AI use cases for your organisation. How would you evaluate them against the six dimensions in Fig. 9.1?
- Which strategy execution tactics do you think would be most successful in your organisation?

Top tip: Using AI to develop AI strategy

Some of the most widely publicised success stories of generative AI have been the use of AI by management consultants. This is not surprising. Consultants advise organisations on strategy development and execution, and a lot of the tasks needed to come up with a good AI strategy can be ably assisted by generative AI.

AI tools like ChatGPT, Claude and Gemini are really good at helping you to frame your thoughts, consider alternative possibilities and evaluate solutions. These are exactly the kinds of skills you need to come up with a good strategy, so consider using AI tools yourself as a key part of developing your AI strategy.

I suggest first giving your favourite generative AI tool a short description of your business and what it is trying to achieve. You can also ask the AI to examine publicly available documents like your annual report or corporate plan. Once you've done this, here are a few prompts you can try out:

> *Prompt: Based on our corporate plan, suggest which of our strategic goals would be best supported by AI. Output should be a list of goals, the problems they address and concrete AI use cases.*

> *Prompt: For each of these use cases, evaluate them against the six dimensions given what you know about my organisation. If there is something you need to know about my organisation to make an assessment, ask me a question rather than making assumptions.*

> *Prompt: I'm trying to decide how best to execute an AI strategy in my organisation. Ask me questions about the organisational culture and once you have enough information, suggest an approach.*

When it comes to strategy, AI is best used as a brainstorming assistant. It isn't yet ready to be a strategy robot and write the strategy for you (Fig. 9.2). But, even if it was, I wouldn't recommend that anyway. Strategy needs to be owned by those who will live and breathe it. And the only way to achieve this is to get the humans involved!

Fig. 9.2. Strategy robot (image generated by ChatGPT).

10

How should I structure my organisation for AI success?

A GOOD AI STRATEGY IS the cornerstone for success with AI. This chapter looks at who will deliver that strategy and how you can put structures in place to ensure good outcomes. This includes roles and responsibilities, organisational structures, and people and cultural aspects that support the strategy.

ROLES AND RESPONSIBILITIES

Who is in charge of your AI strategy and who is ultimately accountable? The answer depends on a lot of factors, including the size of your organisation and AI maturity, as well as your strategic posture.

A growing trend is for companies to create new AI executive roles, often called Chief AI Officers (CAIOs) or an equivalent title. CAIOs take overall responsibility for AI strategy and its implementation, ethics and legal compliance, and change management. The creation of a specific executive role for AI speaks to the value and importance of AI to organisations and the need for a senior person to coordinate across all business and enterprise functions. A December 2024 study of 35,000 public and private institutions in the United States found a 70 per cent year on year increase in the number of CAIO or equivalent roles.[91]

But the CAIO is not the only role that may be given responsibility for AI. Many companies are deciding to rely on existing roles, such as Chief Digital Officers (CDOs), CIOs, or Chief Technology Officers (CTOs). No one choice is better than another, so when deciding, it's a good idea to ask yourself a series of structured questions.

What is my strategic posture?

The previous chapter covered the three AI strategic postures: AI disruptor, AI adaptor, and AI watcher. Whichever of these you choose has a direct bearing on who you delegate to implement the AI strategy. As a disruptor, it is strongly recommended to create a new CAIO role with a direct reporting line to the CEO. You'll need a CAIO who is a disruptor themself and you'll need to empower the CAIO with a high level of authority to get things done. On the other hand, if you are an AI watcher, it doesn't make sense to create a CAIO role. You are likely to have existing roles such as CDO or CIO who can take on the extra responsibility of making incremental decisions about AI and tracking developments.

What strategic plans are already in place?

If you are a relatively mature data-led organisation you will already have data and/or digital strategies in place. You'll have existing teams working on bringing these strategies to life and existing cultural mechanisms to help you roll them out across the organisation. In this case, your AI strategy needs to be strongly aligned to existing plans and shouldn't be seen as a separate activity leading to duplication and confusion. In this case, you can have the CDO take on AI responsibilities, leaving open the possibility of appointing a CAIO later. If you decide to have both a CDO and a CAIO, you should consider reporting lines. A CDO can report to the CAIO if AI is driving business strategy (e.g. AI disruptors), or the CAIO can report to the CDO if proper data management is more important than AI applications. The CDO and CAIO can also be peers reporting to the CIO, COO or CEO.

What is the focus of your AI strategy?

AI strategies can focus on different aspects. Some strategies emphasise technical aspects and require an AI leader with strong technical skills who can build and lead technical teams. In this case, the CTO could be made responsible for the strategy. In other cases, the AI strategy might focus on cultural change, requiring a strong people leader who can drive workforce transformation and assuage AI fears. In such a situation, it might even make sense to delegate AI strategy to the Chief People Officer, although this would be unusual. The previous chapter covered different tactics for executing an AI strategy. Different tactical choices influence who takes responsibility for AI. For example, a centralised approach means a more rigid structure, well suited to a senior CAIO reporting to the CEO. A decentralised approach puts responsibility mainly in the hands of business unit leaders and so there is more collective accountability. In this case, the CIO can

Top tip: Skills of a Chief AI Officer (CAIO)

What skills should a CAIO have? Should a CAIO be a technologist, strategist or a people leader? I argue that a CAIO should be a leader with all three of these skills. A CAIO needs a deep understanding of AI technology and how it's evolving but they need to be equally skilled as a strategic business thinker and have a strong grounding in ethics and governance. Those three skill sets aren't often found together, so you have a challenge on your hands to identify the most appropriate person!

take overall responsibility for AI, acting more in a coordinating capacity.

As you can see, there isn't a single right answer when it comes to determining who takes responsibility for AI strategy. My general guidance is that you need a senior executive with the presence and influencing skills to drive transformation. It's a good idea to avoid leaders with narrow disciplinary experience as AI requires a nuanced understanding of business, culture and technology.

ORGANISATIONAL STRUCTURES

Now you have an AI leader, you will need to set them up with an optimal chance to succeed, which means thinking how your organisation needs to be structured to support execution of your AI strategy. By now, you won't be surprised if I tell you that there's no single right answer. How you structure yourself will depend on your sector, your AI maturity, existing structures, your choice of AI leadership role and your culture.

The good news is that at a high level of abstraction, there are a limited number of choices – centralised, decentralised, hybrid and stand-alone (Fig. 10.1).

	Centralised	Decentralised	Hybrid	AI Business Unit
	CEO, CAIO	CEO, BU, BU, AI, AI	CEO, CAIO, BU, BU, AI, AI	CEO, BU, BU, AI BU
AI maturity	Beginner	Advanced	Intermediate	AI-first
Governance needs	High	Low	Medium	Medium
Business alignment	Medium	High	High	High
Scalability	Medium	High	High	High
Speed of AI deployment	Slow	Fast	Medium	Medium-Fast

Fig. 10.1. Organisational structures for executing AI strategy.

Centralised structure

A common approach is to create a central AI team, often structured as a centre of excellence. In this model, a dedicated AI function – typically led by a Chief AI Officer – oversees AI strategy and execution across the organisation. This model lends itself to highly regulated industries, requiring close oversight of customer impacts and ethical implications. Healthcare is a typical example. The centralised model is often employed by organisations with limited AI experience. In this case, the centre of excellence can act as central point for advice, training and coordination.

Decentralised structure

At the other end of the spectrum is a decentralised model, where AI talent and expertise sit within individual business units. Here, AI is deeply embedded in specific functions such as marketing, operations or customer experience, allowing teams to develop AI solutions somewhat independently. This model enables faster decision-making and experimentation, but it can also lead to silos, inefficiencies and inconsistent governance.

Hybrid

A popular compromise is the hybrid model, which balances central oversight with business-driven AI innovation. A central AI function defines AI standards, governance and best practices, while embedded AI teams in business units drive implementation and tailor AI to their specific needs. This approach avoids duplication of effort while ensuring AI is deployed responsibly and efficiently.

Independent AI business unit

Some companies – particularly those where AI is a revenue driver rather than an internal efficiency tool – go even further, creating an independent AI business unit. In this model, AI is not just an enabler of business strategy, it *is* the business strategy. AI-powered products, platforms or services become core offerings, and the AI function operates as a stand-alone entity, sometimes with its own profit and loss (P&L) account.

The best structure depends on the role AI plays in your organisation. If AI is primarily about efficiency and automation, a centralised or hybrid model may be best. A centralised team will have limited power to disrupt your existing business models. If AI is a competitive differentiator, embedding AI within business units could work better. And if AI is your core business model, treating it as a stand-alone division might make the most sense.

PEOPLE AND CULTURAL ASPECTS

You can have the best AI strategy in the world and it gets you nowhere. You can hire the greatest CAIO ever to walk the earth, and yet still, nothing happens. You can set up organisational structures until you are blue in the face, but it doesn't seem to fix the problem. All you've done is spend a lot of time and effort in planning and organising, but you have nothing to show for it.

This kind of scenario is one I see often. So why does it go wrong?

Any leader knows that a strategy lives or dies on the battlefield of culture. At the end of the day, if your people don't want it, you won't get it. Period.

There's a lot to say about culture and strategy, and I don't have space to go into everything here. What I can do is to give you a few tactics that can help embed your AI strategy in your business. This is not an exhaustive list – and the relationship between culture and strategy can't be boiled down to a checklist. Nevertheless, here are some of the things you should be thinking about to bring your people along for the ride. All the usual things you would do to roll out a new strategy remain relevant. For AI, however, there are additional barriers you will need to overcome.

Training

Let's start with the obvious. To introduce any new technology across your organisation, you will need to accompany the roll-out with simple and effective training. The keywords are *simple* and *effective*. AI can scare people, so you need training that doesn't get bogged down in too many technical details but instead focuses, at least at first, on the value AI can bring to the individual and to the enterprise. Unfortunately, given that AI is an emerging technology, it's not trivial to find simple and effective training courses that work for your organisation. But it is essential. For generative AI, because it seems so intuitive to use, organisations often think training isn't necessary. But as a wide-ranging Australian government study of Microsoft Copilot showed, employees that were trained properly achieved much greater value from the product.[92] The investment required for proper training can be significant, so you will need to factor this into your thinking.

Building trust

Trust is important in any strategy execution and it is doubly important for AI. Globally, trust in AI is low, with 60 per cent of people ambivalent about or distrustful of AI.[93] Employees may fear for their jobs or may simply not trust a technology they don't understand. Reports of AI hallucinations and inherent ethical issues further erode trust. It's a good idea to run surveys in your organisation so you can monitor trust in AI as you roll out your strategy. Building trust takes time but there are concrete things you can do. First, visible AI leadership goes a long way. If your executives are seen to be using AI as part of their work, it gives permission for staff to follow suit. Second, be clear about why you are adopting AI. You may not be able to, but if you can, provide guarantees that there won't be job losses. Third, show how AI is working. Communicate early wins. The more people can see a success story, the more they want to be part of it. Fourth, show when AI is not working. Own it. Communicate that you're aware of the ethical risks and have mitigations in place. But if something still goes wrong, be a learning organisation that actively seeks to do better next time.

Community

You can only drive a strategy top down for so long. At some point, the organisation itself must take it on and run with it. This is where most strategies flounder. After an initial burst of energy and investment, the attention of executives moves on to something else and all that momentum is lost. The antidote to this is to get your community of employees, customers and other stakeholders to own the strategy. That's easier said than done, of course, but encourage communities of practice within the organisation. Seek out AI champions and influencers who naturally bring people together to experiment. Don't try to direct what they do but work with them to incentivise participation. There's no secret sauce here. It is a case of repetitively trying out small nudges until they start to build on each other, layer by layer.

STRATEGY EXECUTION HAPPENS IN WAVES

Implementing a strategy takes time.

I am lucky enough to live in Melbourne, Australia, 200 metres from Port Phillip Bay. When I get stuck on a project, I walk down to the beach and just stare at the water. Even on a calm day, there's a sense of clarity that comes with watching waves relentlessly come ashore.

This is how you want your strategy to be executed. It will happen in waves – waves of excitement and energy inevitably followed by troughs of disappointment as things take longer than expected. But if you persevere, the next wave will come. Be relentless.

Gentle waves are best. Storm waves are magnificent and make a big impression very quickly, but they damage the beach along the way. Think of strategy execution like a series of gentle recurring waves. Eventually, the waves will take on a life of their own. You'll be able to look back and although it seemed like progress was slow at the time, you'll marvel at how far you've come.

QUESTIONS

- Evaluate your organisation against the criteria in Fig. 10.1. Which organisational structure fits your situation best?
- What are the biggest people and cultural challenges you will have when executing your AI strategy?

11

What is the role of my board in adopting AI?

IN THE SCIENCE FICTION SERIES *Severance*, employees at the biotech firm Lumon Industries undergo a radical procedure that separates their work selves from their non-work selves. Each version lives in isolation, one trapped in endless office hours, the other with no memory of work. It's a fascinating take on work–life balance, offering a satirical, sometimes chilling commentary on corporate culture.

For me, some of the most poignant (and darkly amusing) moments come when managers at Lumon interact with 'the board'. Lumon's board is humorously represented by a 1970s-style intercom emitting crackling white noise. For those of us who've dealt with boards, the analogy is painfully familiar. Board directors can sometimes seem like distant, disembodied figures – well meaning, but disconnected from the day to day, their guidance arriving like static on the line.

Yet when it comes to AI adoption, the board's role is anything but abstract. Getting AI right is critical. If things go wrong, the ultimate accountability rests squarely with the board. The challenge is that AI literacy remains low in many boardrooms. As a leader, it is part of your job to close that gap: to help your board understand that AI, like cybersecurity before it, is now a board-level imperative.

Cybersecurity, after all, didn't always command the attention it does today. Cybersecurity now consistently tops the list of concerns that keep board directors up at night,[94] but before the 2000s, it was largely viewed as an IT issue, something for the tech team to worry about. The rise of the internet, and early threats like viruses and malware, gradually brought it into focus. Still, it was treated mainly as a compliance issue until the 2010s, when a surge of high-profile cyber attacks forced boards to take it seriously as a core business risk. Fast forward to today, and most boards have mature cybersecurity practices in place. Directors have become cyber savvy – even those without technical backgrounds.

We are beginning to see a similar shift in how boards engage with AI. AI is quickly climbing the list of issues that preoccupy directors. However, much like the early days of cybersecurity, AI is unlikely to become a true board-level imperative until those occasional sleepless nights give way to full-blown nightmares.

YOUR ROLE AS A LEADER

Whether you're a board director or part of the executive team, you have a role to play in helping your board get ahead of AI. And if, as this book assumes, you're involved in driving AI adoption, raising board awareness should be one of your key objectives. It is far better for the board to understand the risks and implications now rather than discovering them in the headlines further down the track.

What can you do to help your board build maturity around AI adoption? While AI governance is a complex and evolving area, here are three simple steps you can take to get started.

Step one – Educate your board

AI awareness at the board level remains surprisingly low. A 2024 Deloitte global survey of board directors found that for 50 per cent of boards, AI is not even on the agenda.[95] Of all respondents, 80 per cent said their boards had limited or no knowledge or experience with AI. Since the release of ChatGPT, directors, like everyone else, have been inundated with AI media coverage and many are experimenting with generative AI themselves, yet I'm often surprised how many directors I speak to still lack a basic understanding of AI and its evolution.

As a leader, you can play a critical role in shaping your board's engagement with AI. Unless your organisation is already adopting AI on a large scale – at which point it *must* be on the board's agenda – the real challenge is carving out space to educate directors, all of whom are already pulled in multiple directions. A practical starting point is to schedule short, focused speaker sessions during regular board meetings. These might run for just an hour or two but can have a lasting impact.

Speakers can come from within your organisation if someone has the right expertise and board-level presence, but it is often more effective to bring in an external speaker. Avoid inviting individuals with a vested interest, such as tech vendors or consultants pushing their own solutions. Instead, tap into your network to find someone who can offer an independent, balanced perspective.

The best way to build understanding of AI is to start using it. If your directors aren't already using generative AI, or aren't getting the most out of it, consider incorporating a hands-on tutorial into a board session. Tailor the content to typical board activities: reviewing board papers, identifying risks, drafting questions or validating decisions. A simple but powerful exercise is to ask directors to use a generative AI tool to summarise a board paper and discuss whether the output is useful and under what conditions it might (or might not) be appropriate.

The aim isn't to automate board preparation but to spark informed discussion about the practical use and limits of AI in governance. Just be sure to use AI tools with appropriate privacy safeguards so sensitive information isn't exposed to public training systems.

A more radical extension of this idea is to treat AI as a kind of proxy board member. For key decisions, feed relevant information into an AI model and ask for its recommendation. Then compare it with the board's own position. Again, the goal isn't to outsource decision-making but to use AI as a strategic foil, a kind of boardroom devil's advocate that helps surface blind spots and sharpen thinking.

Step two – Audit AI within your organisation

You can't govern what you don't know.

If you want your board to take AI oversight seriously, you need to provide visibility into how AI is being used across the organisation. In other words, a good first step is to conduct an AI audit.

Depending on the size and complexity of your organisation, this may be a straightforward task or a deeply involved one. The effort required also varies by sector. Some industries may use only off-the-shelf generative AI tools for knowledge work, while others deploy more complex, bespoke AI systems for automation, prediction or decision-making.

Take, for example, an aged care provider. Their use of AI may be limited to generative tools, but that doesn't mean it is low risk. You may find staff using generative AI for tasks like writing handover notes between nursing shifts. Is your board aware of these practices? Are they comfortable with them? By contrast, in a large engineering firm, AI use is likely to be more diverse, with business units developing and deploying traditional AI systems for everything from predictive maintenance to operational optimisation. Is the board aware of these efforts? Is there a case for greater standardisation or governance across units?

One challenge in auditing AI use is that employees may not always disclose it, especially if they're using AI tools unofficially. This phenomenon, often referred to as *shadow AI*, involves the unsanctioned use of AI tools without formal oversight. It is a growing issue: one-third of all employees admit to uploading sensitive company data into generative AI systems. Shadow AI

Stakeholder Interviews & Surveys Speak with executives, department heads, IT teams, and end-users to uncover current AI use.	✓ or X
Application & System Inventory Catalogue commercial software, subscriptions, and in-house AI tools—including any shadow AI.	✓ or X
Workflow & Process Mapping Identify where automation or AI-supported decision-making is embedded in operational processes.	✓ or X
Spend & Vendor Analysis Track AI-related expenditures through budgets, vendor contracts, and cloud service usage.	✓ or X
Sentiment & Adoption Assessment Survey employees to understand how AI is perceived and used in practice.	✓ or X

Fig. 11.1. An AI audit checklist.

not only makes auditing more difficult but it also introduces real security and compliance risks.

To bring structure and consistency to your AI audit, consider using a checklist of activities that can be shared with your board (Fig. 11.1).

It is good practice to repeat AI audits periodically and report findings to the board. This not only provides oversight but also helps track how AI use is evolving over time. You might also consider assessing your organisation's progress using an AI maturity framework. Many such frameworks exist (e.g. from Gartner, Boston Consulting Group, IBM) and can help clarify how systematic your AI adoption is. Knowing where you currently stand on a maturity model can be a powerful way to spark a strategic conversation with your board: where do you want to be and what's the plan to get there?

Step three – Managing AI risk

Boards must evolve their risk management practices for the age of AI.

One of the core responsibilities of any board is to understand and manage risk. As AI adoption accelerates, it introduces new categories of risk, ranging from strategic and operational to ethical

Table 11.1. Categories of AI risk for boards.

Risk category	Definition	Examples
Ethical risks	Risks related to fairness, accountability, bias and transparency in AI use.	Discriminatory hiring algorithms, opaque decision-making, lack of explainability.
Regulatory risks	Risks of non-compliance with existing or emerging AI-related regulations.	Violating GDPR through AI-driven profiling; non-compliance with the EU AI Act (2024).
Operational risks	Risks from failures in AI systems that affect operations or service delivery.	Model drift, poor training data, unpredictable outcomes, failure in automation processes.
Cybersecurity risks	Risks involving the security of AI systems and AI-enhanced cyber threats.	Adversarial attacks, data poisoning, AI-generated phishing, model hijacking.
Reputational risks	Risks of negative public perception or erosion of stakeholder trust.	AI misuse scandals, biased chatbot responses, customer backlash over unethical AI practices.
Strategic risks	Risks related to AI's impact on the organisation's business model or position.	Falling behind AI-enabled competitors; misaligned AI investments; disruption of core markets.

and environmental. These include risks like losing market share to AI-enabled competitors, workforce disruption from automation, regulatory and legal exposure and growing scrutiny of AI's environmental and social impact (Table 11.1). Part III of this book provides a deep dive into these social, environmental, and ethical risks. Here we focus on how boards can adapt their existing risk management frameworks to include AI.

A separate AI risk framework? Not necessary

My main advice for boards is this: you don't need to create an entirely new risk framework for AI. A well-designed enterprise risk management framework already accounts for a wide range of risks: financial, legal and regulatory, reputational, workforce, infrastructure (including IT) and cybersecurity. AI risk can and should be integrated into these existing categories.

That said, some aspects of AI may require specialised governance mechanisms, just as cybersecurity and environmental compliance do, but the goal should be integration, not isolation. Hence, to account properly for new risks from AI adoption, a good approach is to review your organisation's risk appetite, risk management structures and policies and procedures.

Define risk appetite for AI

Boards are responsible for setting the organisation's overall risk appetite, determining the level of risk acceptable in pursuit of strategic goals. Most organisations already have a risk appetite statement. If yours doesn't, now is the time to create one.

When it comes to AI, boards should assess risk tolerance across the six categories in Table 11.1. Risk tolerance is usually categorised as low, medium or high. For instance, a government agency may have low tolerance for ethical and reputational risks, while a fast-moving tech startup might accept more reputational risk in pursuit of innovation. Likewise, a traditional bank is likely to maintain a low tolerance for AI-related cybersecurity threats, whereas a crypto startup might have a greater tolerance.

Importantly, risk appetite should align with your organisation's strategic posture towards AI. Using AI to disrupt your industry often entails greater risk. Boards should ensure their risk appetite supports rather than conflicts with the ambitions in their AI strategy.

Establish risk management structures

Once the risk appetite is defined, the board must ensure proper oversight structures are in place. These should be integrated into the organisation's broader governance system:

- **Risk committees.** Most boards already have Audit and Risk Committees. These are often well placed to assume responsibility for AI risk. Depending on maturity and posture, some organisations establish AI-specific committees, like an AI Ethics Committee, or bring in external advisory boards to address gaps in expertise.

- **Chief Risk Officers (CROs)**: The CRO plays a central role in keeping the organisation within its risk appetite. The board should assess whether existing roles (e.g. compliance officers, cybersecurity risk leads) can manage AI risks, or if a dedicated AI risk officer is needed – particularly in larger or more AI-intensive organisations. This role should work closely with the CDO, CIO and the Chief Information Security Officer (CISO) where applicable.
- **External reporting**: Boards already have statutory duties around financial and audit compliance. While no country currently mandates AI-specific disclosures, governance codes in France and Germany now require directors to educate themselves on emerging technologies. Globally, over 1,000 AI policy initiatives have been proposed across 69 countries.[96] The EU AI Act, enacted in 2024, was the first comprehensive AI regulation, taking a risk-based approach and banning certain applications. Regulatory divergence is also emerging: for example, President Trump's 2025 administration reversed Biden-era executive orders on AI governance, opting for a deregulated approach.

Review and update policies and procedures

Boards must ensure organisational policies effectively mitigate AI-related risks without becoming overly bureaucratic or stifling innovation. In most cases, you will need at least one new AI policy. Even basic use of generative AI tools requires guidance on appropriate use: for instance, prohibiting staff from uploading sensitive data.

For broader AI adoption, you are likely to need a more comprehensive set of policies addressing:

- risk ownership and mitigation
- ethical principles and AI use boundaries
- testing and validation before deployment
- incident response
- staff training and awareness
- policy enforcement and monitoring.

The Corporate Governance Institute recommends a 10-step process for developing an AI policy, covering everything from identifying use cases and risks to establishing accountability and monitoring.[97] These steps align closely with the content in this book: AI objectives and use cases are addressed in your strategy; roles and responsibilities are covered in the organisational structure chapter; and governance and risk management are the focus here.

SUMMARY OF BOARD ROLES FOR AI

Figure 11.2 summarises the risk roles of a board and how these should be reviewed in the light of AI adoption. Your board

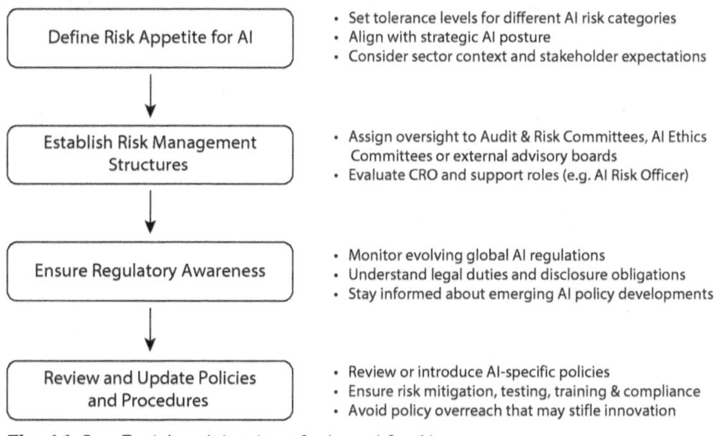

Fig. 11.2. Evolving risk roles of a board for AI.

Top tip: Further resources

The Australian Institute of Company Directors (AICD) has a comprehensive and accessible guide on AI for directors. It covers practical steps on roles and responsibilities, governance structures, policies and strategy, culture, monitoring and evaluation.[98]

For more information on emerging AI regulation around the world, the OECD AI Observatory maintains a website with up-to-date resources on AI policies, regulations and strategies in over 70 countries.[99]

should now be well equipped to adopt AI in a considered and responsible way.

QUESTIONS

- On a scale of 1 to 10, how mature is your board when it comes to AI education and awareness?
- Think about your organisation's current risk appetite. Would AI adoption change anything?
- Does your organisation have any AI policies in place? What is missing?

Part IV
The bigger picture

The bigger picture (image generated by ChatGPT).

This section of this book turns to the *bigger picture* of AI, issues that extend beyond any single use case or application. While the previous section focused on risk and governance, the chapters that follow explore four broad themes that should also be integrated into your organisation's overall risk management approach:

- *AI ethics.* This chapter presents a thorough and systematic approach to the ethics of AI and how to adopt AI in a responsible way.
- *AI and environmental impact.* AI, and the data centres that run it, is having a significant impact on the environment. As the scale of AI increases rapidly, the industry places strains on energy and water use as well as demands for the critical minerals that devices that run AI use.
- *Environmental, social and governance (ESG).* This chapter delves deeper into the potential environmental and social impacts of AI and addresses how to include AI-related issues in ESG monitoring and reporting frameworks.
- *AI and cybersecurity.* AI technologies can help prevent cyber attacks. On the other hand, AI can lead to new threats because of the way that AI systems are adopted and used. This chapter covers both sides of this coin to give you a deep understanding of how AI and cybersecurity interact.

12

What are the ethics of using AI in my organisation?

ORGANISATIONS ARE INCREASINGLY DRIVEN BY social responsibility, not just profitability. Many countries now have legislation that mandates disclosures related to ethical practices, social impact and sustainability. For example, the UK's Modern Slavery Act requires companies of a certain size to disclose efforts to prevent slavery in their supply chains. In the United States, the Dodd–Frank Act requires disclosure of critical minerals sourcing to ensure a given company's supply chain of these does not support armed groups in conflict zones. The EU's General Data Protection Regulation (GDPR) imposes hefty fines on companies found to breach transparent data processing practices.

When adopting AI in your organisation, you should think about how your use of AI affects your social responsibility goals. AI can provide positive benefits to society, helping with environmental sustainability, human rights and inequality. But it can also have very serious negative consequences.

There are thousands of examples of how AI is being used in a positive way for society. AI is helping to preserve the Great Barrier Reef,[100] using underwater gliders with computer vision capabilities to improve monitoring of the growth of invasive species. AI is fighting wildfires, providing better predictions of how fires spread so that resources can be assigned more effectively.[101] AI is reducing waiting times and hence patient anxiety in breast cancer scans.[102] It is helping with speech development for those with autism and Parkinson's disease.[103]

Like any technology, AI has its downsides too. The rapidly increasing energy and water demands of data centres is contributing to global warming. AI-driven social media algorithms are leading to a rise in mental health issues in teenagers.[104] And AI can increase social exclusion, resulting in an 'AI divide' between high-income and low-income countries.

It is critical, therefore, that any business adopting AI carefully considers both the positives and negatives of its use of AI on society. There are many frameworks to help you think about this; the problem comes down to whether your application of AI can be considered ethical, from a social and environmental point of view.

Top tip: Is it ethical?

Between 2018 and 2021, Bunnings, a major hardware chain in Australia and New Zealand, trialled the use of AI-powered facial recognition in 63 of its stores. The initiative aimed to protect staff and customers following a rise in violent and aggressive in-store incidents. But that wasn't how the public perceived it. The trial triggered widespread backlash as it became clear that sensitive biometric data was being collected without customers' consent or even notification.

Was the use of AI ethical in this case? Does the duty to provide a safe and secure environment for customers and employees justify such surveillance? Or should facial recognition be off limits entirely in retail contexts? And perhaps most importantly, could Bunnings have taken steps to address the ethical concerns through greater transparency, stronger consent mechanisms or tighter data controls?

THE ETHICS OF AI

The term *ethics of AI* refers to the practice of designing and implementing AI systems in ways that align with broadly accepted moral standards. Because AI automates decision-making, it often raises complex ethical dilemmas, particularly when its actions may lead to negative or unintended consequences.

One of the most well-known ethical thought experiments is the *trolley problem*, which explores the morality of sacrificing one life to save many. In the context of AI, a common variation involves self-driving cars. Imagine an autonomous vehicle is about to crash and there is no option that avoids loss of life. Should it stay on course and hit a group of pedestrians, or swerve and endanger the passengers inside? Which choice should the AI make? And who gets to decide?

The trolley problem is an extreme case of an AI ethical dilemma, and you're unlikely to encounter it when adopting AI in your own organisation. However, almost any application of AI needs to consider ethical issues. Should a law enforcement agency use facial recognition technology to hunt for criminals, even if it violates ordinary citizens' privacy? Should a healthcare company use AI-driven remote monitoring if it can save lives but disadvantages patients who can't afford the technology? Should a retail chain use AI to drive sales of unhealthy food?

Fortunately, there has been substantial progress in developing guidance for ethical AI adoption. For instance, Table 12.1 outlines the Australian Government's AI Ethics Principles. These are high-level guidelines designed to support organisations in using AI responsibly. Similar frameworks exist in many countries. The OECD's AI Policy Observatory maintains a global inventory of national-level initiatives aimed at promoting ethical AI practices.[100]

While these principles began as voluntary, recent years have seen a shift towards formal regulation. The most significant

Table 12.1. Australia's AI ethics principles.

Principle	Description
Human, societal and environmental well-being	AI systems should benefit individuals, society and the environment.
Human-centred values	AI systems should respect human rights, diversity and the autonomy of individuals.
Fairness	AI systems should be inclusive and accessible, and should not involve or result in unfair discrimination against individuals, communities or groups.
Privacy protection and security	AI systems should respect and uphold privacy rights and data protection, and ensure the security of data.
Reliability and safety	AI systems should reliably operate in accordance with their intended purpose.
Transparency and explainability	There should be transparency and responsible disclosure so people can understand when they are being impacted by AI, and they can find out when an AI system is engaging with them.
Contestability	When an AI system significantly impacts a person, community, group or environment, there should be a timely process to allow people to challenge the use or outcomes of the AI system.
Accountability	People responsible for the phases of the AI system life cycle should be identifiable and accountable for the outcomes, and human oversight of AI systems should be enabled.

example is the EU AI Act,[105] now in force, which marks a turning point in how AI is governed globally.

AI ETHICS AND VALUES

Ethics can be a challenging concept to navigate. Ethical norms differ across individuals, organisations and cultures. And ethical guidelines are typically broad and abstract, making them difficult to make operational or measure in practice.

For organisations working through the ethical implications of AI, I recommend shifting focus towards the related, and often more practical concept of *values*. The distinction between ethics and values is subtle but important. Ethics are societal norms that guide behaviour and carry a specific moral weight. Values, on the other hand, are the things individuals or organisations care about. They don't necessarily have a moral dimension but they reflect what matters most to a given group.

Focusing on *values* rather than *ethics* offers two key advantages. First, the concept of values tends to be less loaded and more practical than ethics. Values can often be defined more precisely and, importantly, measured. Ethics, by contrast, can feel vague or abstract, leading to confusion and inconsistency in implementation.

Second, many organisations already have a set of corporate values. By anchoring your approach to AI in these existing values, you can build on established cultural foundations and align your AI efforts with your broader organisational identity.

That said, a word of caution: corporate values statements typically reflect an organisation's *internal* priorities, not a comprehensive framework for anticipating the *external* risks and societal implications of AI. It's wise to use your corporate values as a starting point, but don't let them limit your thinking.

A THREE-STEP PROCESS FOR ETHICAL AI DECISION-MAKING

Given the ambiguity often associated with ethical principles, adopting a structured approach is the most practical way to navigate the ethical implications of AI. While many public frameworks are available,[106] these are often lengthy, overly detailed, or difficult to make operational.

To simplify the complexity, this chapter introduces a clear, three-step process (Fig. 12.1). The process begins with clarifying your values, moves to implementing practical mitigations and concludes with continuous improvement.

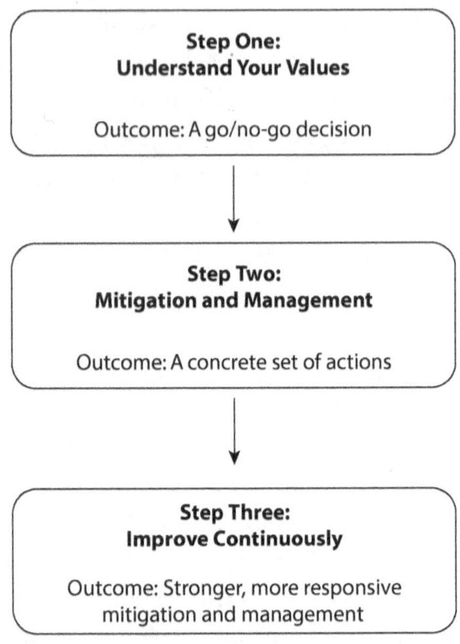

Fig. 12.1. A three-step process for ethical AI decision-making.

Step one: Understand your values

Outcome: A go/no-go decision

The first and most important step is to clarify your organisation's values as well as those of your key stakeholders. Without this clarity, it is difficult to judge whether a proposed AI use case aligns with your organisation's identity and obligations.

Start by reviewing or establishing a set of priority values. Table 12.2 offers a comprehensive set of common values, grouped by category, as a starting point. You don't need to adopt them all. Choose the ones most relevant to your organisation and context. If you already have a corporate values statement, use that as a foundation but be open to expanding it to reflect the broader implications of AI.

Table 12.2. Common values for ethical AI assessment.

Category	Values
Core values	Honesty, Integrity, Respect, Fairness, Accountability, Compassion, Trustworthiness
Relational and social	Empathy, Collaboration, Kindness, Inclusivity, Tolerance
Civic and community	Equality, Civic responsibility, Environmental stewardship, Solidarity
Professional and organisational	Excellence, Innovation, Transparency, Impartiality, Diligence
Human-centric	Autonomy, Human dignity, Prudence, Courage, Humility
Universal and global	Sustainability, Peace, Non-maleficence (do no harm), Beneficence (do good)
Context-specific	Confidentiality, Cultural sensitivity, Adaptability

Next, assess how your proposed use of AI could impact each of your chosen values. If you have defined an AI strategy, with a set of AI use cases, evaluate each of these use cases against the value set.

For example, if transparency is a core value, consider whether AI decisions can be easily explained to customers affected by them.

If environmental stewardship is key, evaluate the full environmental impact of the AI system, including its energy use and the carbon footprint of your data infrastructure and supply chain.

At this stage, you're not designing specific mitigations. Instead, you're asking the following question: if the right safeguards existed, would this AI use case align with our values? If the answer is yes, proceed to step two. If not, stop.

Step two: Mitigation and management

Outcome: Concrete actions to manage any ethical risks

Having identified the major ethical risks in step one and deciding to proceed, you can now turn to practical mitigation and management strategies.

Table 12.3 outlines commonly used actions you can take, grouped into three categories:

- *Project-level actions*: Practical steps integrated into your AI project life cycle, such as defining measurable ethical requirements or creating mechanisms for human oversight.

Table 12.3. Example mitigations and management practices for ethical AI.

Category	Action	Definition
Project level	Ethics by numbers	Define ethical AI requirements in quantifiable, measurable terms.
	Data blueprint	Clearly articulate ethical data requirements at all stages of the project.
	Ethical playbook	Ensure ethical AI considerations are integrated into project plans.
	Risk rehearsal	Conduct thorough testing and simulations to mitigate unforeseen ethical risks.
	Explain-it-all	Design AI systems to provide clear, understandable explanations of their decisions.
	Compliance tracker	Implement automated data collection to support ethical monitoring and compliance.
	Kill switch	Create failsafe mechanisms to isolate and disable AI components in case of issues.
	Supplier vetting	Introduce processes to ensure suppliers adhere to your ethical AI guidelines.
	Double duty	Build in multiple AI components for critical tasks to enhance reliability.
	Ethics guardian	Assign a team member as a 'Critical Friend' to identify potential ethical violations.
	Ethics radar	Implement real-time monitoring to flag potential ethical issues dynamically.
Organisational	Policy tune-up	Review and update organisational policies to support ethical AI practices.
	Ethics bootcamp	Provide comprehensive training on ethical AI principles for employees.
	Hire for values	Align hiring practices with ethical AI values to ensure cultural fit.

Category	Action	Definition
	Brand of trust	Position the organisation as a leader in ethical AI through transparent practices.
	Values check	Reassess corporate values to ensure alignment with ethical AI goals.
	Ethics hub	Establish communities of practice to foster open discussions on AI ethics.
	Top-down ethics	Have senior leaders actively champion ethical AI initiatives.
	Ethics stars	Recognise and reward employees who exemplify ethical AI principles.
	Safe space	Foster a culture of accountability without placing blame for ethical missteps.
	Ethics council	Create a board or committee to oversee AI ethics across the organisation.
	Transparency pulse	Publish regular reports detailing ethical AI practices and outcomes.
	Speak up!	Establish anonymous channels for raising ethical concerns.
	Culture barometer	Assess the organisation's culture to ensure it supports ethical decision-making.
External	Standards spotlight	Document and communicate alignment with ethical standards.
	Law watch	Stay informed about and comply with relevant AI laws globally.
	Community first	Engage user groups and communities to ensure alignment on ethical AI principles.
	Audit allies	Engage independent auditors to assess ethical compliance of AI systems.
	Open forum	Host public forums to gather input on AI-related ethical concerns.
	Partner values	Collaborate with partners who share similar ethical values and practices.
	Policy shapers	Actively participate in shaping ethical AI regulations and policies.
	Crisis shield	Develop a public response strategy for addressing ethical controversies.
	User school	Provide resources to educate customers on AI use and ethical safeguards.

(Table 12.3 taken from Lu *et al.*[107] See there for a fuller list.)

- *Organisational actions*: Strategies addressing broader organisational culture, training, policies and leadership involvement.
- *External actions*: Practices to manage your external environment, including stakeholder engagement, regulatory compliance, independent audits and public transparency.

You don't need to adopt every listed action. Instead, select actions aligned to your prioritised values from step one. For instance, if environmental stewardship is important and you're using AI to manage your supply chain, you might:

- Implement a *data blueprint* ensuring environmental credentials are included in decision datasets.
- Conduct *supplier vetting* to ensure partners share your environmental values.
- Encourage a *speak-up* culture where employees can raise ethical concerns freely.
- Engage local communities proactively (*community first*).

Step three: Continuous improvement

Outcome: Ongoing refinement of mitigations and management practices

The project-level, organisational and external actions in Table 12.3 are a good way to manage ethical AI risks by actively managing how your AI use cases are implemented. This is a form of good governance. See the earlier chapter on governance for a fuller discussion of this topic.

A key element of governance is continuous monitoring and evaluation. This is a critical step to ensure that your actions have the desired effect. It is even more critical for AI. The statistical nature of AI means that there can be surprises once an AI system is fully deployed. You need to monitor not just the technical performance of the AI system but also how the broader community perceives it. For example, you might implement the perfect AI decision-making system to decide which customers get a mortgage. But if customers perceive bias – whether it exists or not – your reputation is at stake.

Think of your mitigation and management approach as an iterative activity. You won't get it right first time, so review its effectiveness on an ongoing basis and change or add actions as you need. Foster a culture of continuous improvement. This is good practice in any business, but for AI, given its complexity, it is essential.

It is also a good idea to revisit the go/no-go decision periodically, perhaps every 12 months, to assess whether to continue with your

Top tip: Integrating ethical AI into governance

The mitigations outlined in this chapter are essentially a specialised form of governance tailored specifically for managing ethical AI risks. They complement, not replace, the broader AI governance guidance provided earlier in this book.

To manage ethical AI risks effectively:

- Incorporate ethical considerations explicitly into your broader AI governance framework.
- Recognise that ethical risks require deeper analysis and special attention compared to other risk categories.
- Use the governance structures previously described, such as delegated committees and clear accountability, to address ethical concerns at operational levels while maintaining visibility and oversight at the board level.

Ethical AI governance isn't a stand-alone task; it should be deeply embedded in your organisation's overall governance strategy.

AI use case or change course. Be prepared to pull back at any moment. It is almost impossible to predict how the public will perceive your use of AI and if there is serious negative backlash, it pays to be prepared.

QUESTIONS

- Consider the example of the Bunnings facial recognition system. What values should have been important for Bunnings to consider? How was this use of AI misaligned to those values?
- Think of a potential AI use case in your organisation. What values are important for you to consider?
- What project, organisational and external actions could you take to mitigate and manage any potential ethical AI risks?

13

What is the impact of AI on the environment?

It has been claimed that a query on a generative AI chatbot consumes 10 times the energy of the equivalent Google search.[108] A Google search hunts for a list of sources, returns a set of links and lets the user do the work of choosing information from them. A generative AI search, on the other hand, does that work for the user, intelligently analysing a range of sources and generating a summary. Generative AI can do things that a Google search can't, but because it requires more search, more analysis and more text generation, it is also a lot more energy intensive. Generative AI is increasingly seen as an alternative to traditional search, so the energy demands of one of the most common digital tasks is ballooning.

At first glance, the energy use of generative AI search appears to be problematic. On the other hand, if generative AI can synthesise information from websites more efficiently than humans, could it be that the overall energy use – compared to the user clicking on and reading multiple Google links – is actually lower? A 2024 study in *Nature* showed that the carbon emissions of an AI writing a page of text are up to 1,500 times lower than those of a human using a computer to write the same text.[109]

Whether or not AI uses more energy than traditional computing is a bit of a moot question. It is undeniable that the AI industry is a major consumer of power. The cause isn't just users interacting with AI, it's the entire supply chain of building and deploying AI devices across the world in mass numbers. Many users of AI are simply unaware of the implications for the environment. With AI use only likely to grow in the future, it is important to create more transparency about AI's impact and come up with ways to minimise it.

HOW AI IMPACTS THE ENVIRONMENT

The AI supply chain leads to significant negative impacts on the environment. To illustrate this, consider a simple AI-driven web search.

A single search query requires a complex set of activities. The user must interact with a computing device, such as a mobile phone or laptop. Manufacturing these devices requires extracting critical minerals from the earth, and once the life of the device is over, it is most likely that the device becomes e-waste. Then there's the development of the AI model itself, which requires training on vast quantities of data. All this is before the actual web search. The world carries out billions of web searches every day, each one requiring computing resources. Computations take place in the cloud in huge data centres that are significant construction projects, requiring massive amounts of energy to build and operate.

Each step in the AI supply chain has significant environmental impact (see Fig. 13.1), although the exact scale of the impact is not well understood. There are many moving parts so it is difficult to come up with realistic estimates of environmental metrics like energy usage. In her book, *Atlas of AI*, Kate Crawford argues that the tech sector treats such information as 'highly guarded corporate secrets', publicly touting corporate sustainability initiatives whilst, at the same time, avoiding discussion on measurement of environmental impact.[110] Research is now beginning to emerge, however, with one of the first papers in the

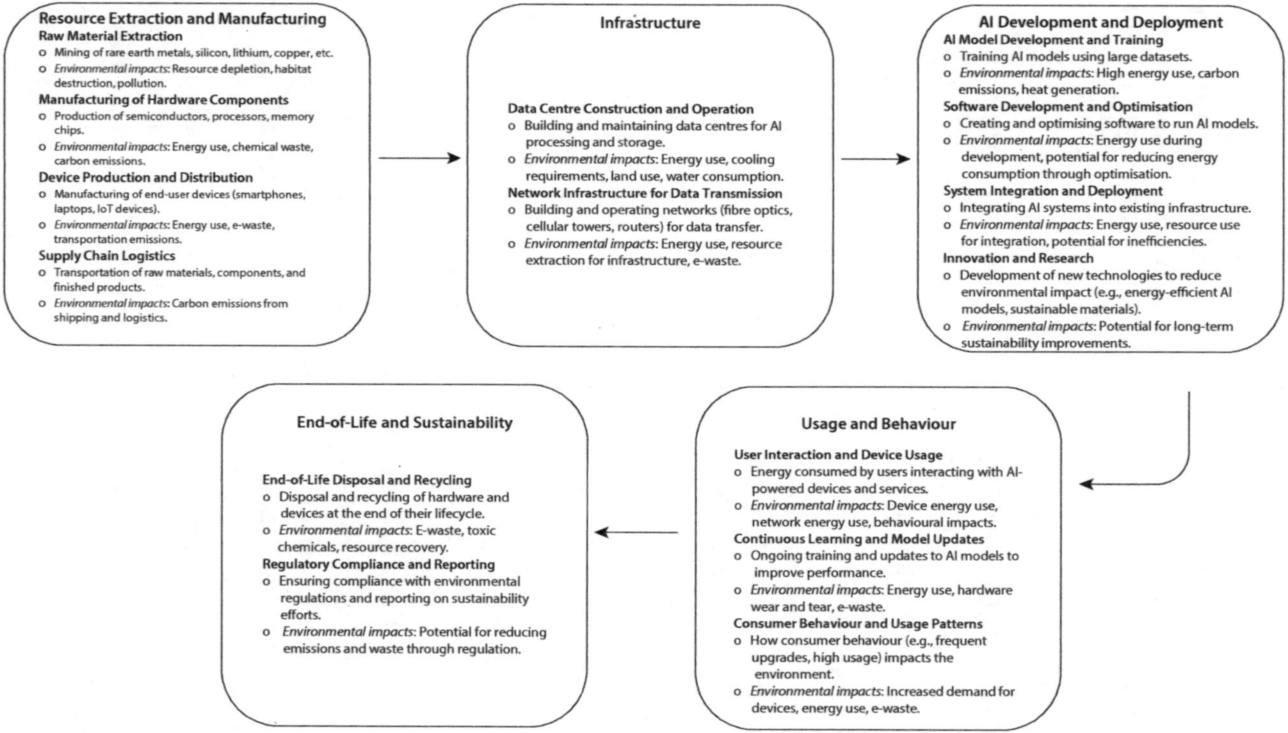

Resource Extraction and Manufacturing

Raw Material Extraction

- Mining of rare earth metals, silicon, lithium, copper, etc.
- *Environmental impacts*: Resource depletion, habitat destruction, pollution.

Manufacturing of Hardware Components

- Production of semiconductors, processors, memory chips.
- *Environmental impacts*: Energy use, chemical waste, carbon emissions.

Device Production and Distribution

- Manufacturing of end-user devices (smartphones, laptops, IoT devices).
- *Environmental impacts*: Energy use, e-waste, transportation emissions.

Supply Chain Logistics

- Transportation of raw materials, components, and finished products.
- *Environmental impacts*: Carbon emissions from shipping and logistics.

Infrastructure

Data Centre Construction and Operation

- Building and maintaining data centres for AI processing and storage.
- *Environmental impacts*: Energy use, cooling requirements, land use, water consumption.

Network Infrastructure for Data Transmission

- Building and operating networks (fibre optics, cellular towers, routers) for data transfer.
- *Environmental impacts*: Energy use, resource extraction for infrastructure, e-waste.

AI Development and Deployment

AI Model Development and Training

- Training AI models using large datasets.
- *Environmental impacts*: High energy use, carbon emissions, heat generation.

Software Development and Optimisation

- Creating and optimising software to run AI models.
- *Environmental impacts*: Energy use during development, potential for reducing energy consumption through optimisation.

System Integration and Deployment

- Integrating AI systems into existing infrastructure.
- *Environmental impacts*: Energy use, resource use for integration, potential for inefficiencies.

Innovation and Research

- Development of new technologies to reduce environmental impact (e.g., energy-efficient AI models, sustainable materials).
- *Environmental impacts*: Potential for long-term sustainability improvements.

Usage and Behaviour

User Interaction and Device Usage

- Energy consumed by users interacting with AI-powered devices and services.
- *Environmental impacts*: Device energy use, network energy use, behavioural impacts.

Continuous Learning and Model Updates

- Ongoing training and updates to AI models to improve performance.
- *Environmental impacts*: Energy use, hardware wear and tear, e-waste.

Consumer Behaviour and Usage Patterns

- How consumer behaviour (e.g., frequent upgrades, high usage) impacts the environment.
- *Environmental impacts*: Increased demand for devices, energy use, e-waste.

End-of-Life and Sustainability

End-of-Life Disposal and Recycling

- Disposal and recycling of hardware and devices at the end of their lifecycle.
- *Environmental impacts*: E-waste, toxic chemicals, resource recovery.

Regulatory Compliance and Reporting

- Ensuring compliance with environmental regulations and reporting on sustainability efforts.
- *Environmental impacts*: Potential for reducing emissions and waste through regulation.

Fig. 13.1. Environmental impacts across the AI life cycle.

field reporting that running a single AI model produced more than 660,000 pounds of carbon dioxide emissions, equivalent to 125 round-trip flights from New York to Beijing.[110] And this was in 2019, three years before the launch of ChatGPT and the beginning of exponential growth in the use of generative AI models.

THE ENVIRONMENTAL LIFE CYCLE OF AI SEARCH

Let's go back to our simple AI-driven web search and look at the environmental impacts across the AI life cycle (Fig. 13.1).

Computing device impacts

First, you need a device to carry out your search. Let's assume you are using a mobile phone. An iPhone contains over 30 critical minerals and rare earth elements.[111] While a significant proportion of these are now recycled – for example, 75 per cent of aluminium in the iPhone 15 is recycled[112] – recycling rates of mobile phones remain low.[113] This means that the tech industry is reliant on large-scale mining of critical metals and rare earth minerals. According to Kate Crawford, the ratio of usable minerals to waste toxins in a mining process is extreme, with up to 99.8 per cent of earth removed ending up as waste, often toxic.[111]

Computational infrastructure impacts

The computational infrastructure – both software and hardware – required to power the billions of web searches that take place every day is significant. Nowadays, most computation is carried out in the cloud, on virtual servers that could be located anywhere in the world, rather than on the user's own local machine. The word 'cloud' conjures up images of white, fluffy objects. As a result, we tend not to think about the very real and physical infrastructure that runs the cloud. This infrastructure consists of massive data centres that house the semiconductors, servers, power and cooling technology necessary to run modern-day computations.

Data centres are gigantic, with some of the largest having the footprint of 25 football fields.[114] Data centres consist of thousands of racks, organised in lines, each containing servers, networking equipment and storage. The largest data centres cost billions of dollars to build.[115] And demand for data centres is continuing to grow, with US demand expected to increase by 10 per cent annually until 2030.[116] The rise of so-called AI data centres is estimated to lead to even further growth, possibly up to a doubling in capacity by 2030. The unique needs of AI require processing massive amounts of data. In simple terms, AI is 'super-sizing' data centres, which are now bigger, denser and more energy-demanding than ever before.

As well as the impacts of building large data centres, there are two key concerns when it comes to their operation: energy and water.

Electricity for data centres

The International Energy Agency (IEA) has estimated that data centres account for 2 per cent of all global electricity usage, with demand expected to grow significantly.[116] The energy demands vary considerably by country. In the United States, home to a third of the world's data centres, power use is 6 per cent of the country's overall consumption, but in Ireland, a relatively small country which houses over 80 data centres, that figure is expected to grow to 32 per cent in the near future.[117]

Some commentators consider 2 per cent of global electricity usage to be modest. Even considering 'fast growth' scenarios for data centres, the IEA still sees data centre demand dwarfed by use from heating and air conditioning, electric vehicles, and other industries. However, it is undeniable that the growth of AI is increasing energy demands. AI GPU chips use roughly twice the amount of energy of traditional computer processing units (CPUs), and Deloitte predicts that if only 5 per cent of internet searches switch to generative AI prompt-based searches, the increase in energy would be equivalent to over 100,000 US households.[117] AI is, however, not the only source of increasing energy consumption in tech. While estimates vary significantly, most sources say that cryptocurrency mining continues to burn more energy than AI.

Water cooling in data centres

Data centres generate a lot of heat, and all that heat needs to be dissipated, either by air conditioning or liquid cooling systems. AI data centres most commonly rely on water-based cooling, meaning that the demand for fresh water in AI data centres could be as high as 1.7 trillion gallons (6.5 trillion litres) by 2027.[118] Shaolei Ren and his team have studied the water needs of AI data centres and give a thorough account of exactly how a data centre uses water.[119] He estimates that 10 to 50 typical prompts to a generative AI tool consumes a 500-millilitre bottle of water.

The world has a water-scarcity problem, not because of a lack of water but because water isn't equally distributed across the globe. Given the high demands for water in AI data centres, there is a risk that the divide between those who have and those who do not have water will increase. Data centres have already led to new problems of water scarcity in drought-ridden areas. There is a data centre in Altoona, Iowa, that consumes one-fifth of the town's water supply, despite Iowa going through one of the most prolonged droughts in its history.[120]

Impact of training AI models

Ever since the onset of large language models (LLMs), the emphasis in the AI world has been on the word 'large', as tech companies have tried to outdo each other to create bigger and bigger AI models. When ChatGPT launched in November 2022, its model had 175 billion parameters (which already dwarfed the 117 million parameters of the first version of the model in 2018).[121] A later version of a model from OpenAI, GPT-4, has over 1.75 trillion parameters.[122] The cost – and environmental impact – of training such large models is enormous. The electricity required to train GPT-3 is estimated at around 1,287,000 kilowatt hours[123] roughly equivalent to two and a half long-haul flights. It is important though to note that a large language model doesn't need to be trained each time it is used. It undergoes a one-time training process, followed by occasional retraining, unlike, say, the aviation industry, where a similar amount of electricity is consumed every time an aircraft takes flight.

Impact of AI inference

Having considered the environmental impact of manufacturing and operating devices that run AI, of building and maintaining AI data centres, and of training AI models, we have finally arrived at the actual web search itself. All that work – devices, data centres, model training – is needed before you can run an AI search in your favourite app or browser. The technical term for asking an AI model to do something useful is *inference*. In our case, the AI search is an example of inference.

How energy-intensive is an AI inference that uses a large language model? This question takes us right back to the beginning of this chapter where I mentioned that an LLM search can consume 10 times more energy than a traditional web search. But what is the carbon footprint of inference compared to model training, and is it significant in the context of the overall footprint of a society?

Studies show that model inference accounts for up to 90 per cent of the costs of deployed AI systems.[124] That is surprising. We have learned that AI needs vast amounts of computation to train models with trillions of parameters. How can a simple search be more energy intensive? It is down, of course, to multiple factors. Google processes 8.5 billion searches every day,[125] and if each one produces an LLM-based search summary, that adds up quickly.

The relative environmental impact of AI

Of course, the real answer about the carbon footprint of AI inference is not so simple. It is unclear if an LLM-based search is more efficient than a traditional search because it reduces the need for the user to carry out multiple searches. And, in the big scheme of things, the energy used by AI inference pales in comparison to how we use energy elsewhere in our daily lives. Add the fact that AI computation is becoming more efficient and there is a trend towards smaller models, and it could be argued that the environmental impact of AI is comparatively modest. That is not to ignore the very real impacts. But, as with all things AI, it is never as simple as black and white. We do need more effort to reduce the carbon footprint of AI, but, at the same time, we shouldn't neglect other high energy-consuming parts of the economy.

TOWARDS A MORE SUSTAINABLE AI INDUSTRY

There are active efforts in the industry aimed at reducing the environmental impact of AI. For data centres, the trend is towards more sustainable ways of building and operating them. The efficiency of AI computer chips housed in data centres is improving exponentially, and the industry is aggressively moving towards more use of renewable energy sources.

Progress varies across geographies and organisations. IBM has claimed that 74 per cent of energy use in its data centres comes from renewable sources.[126] While some data centre operators are building their own on-site renewable energy generation systems, most data centres are powered from the local grid, meaning that the energy mix is dependent on the energy sources used in that location.[127] With 80 per cent of energy in the United States still coming from fossil fuels, this inevitably means that currently, most data centres do not run on renewable energy. However, tech companies are investing heavily in direct sourcing of renewable energy, as well as alternative sources such as hydrogen and nuclear. Google says it will run completely on carbon-free energy by 2030.

When it comes to water demands, the industry is again experimenting with more sustainable solutions. A novel groundwater cooling system now operates at the Pawsey Supercomputer facility in Perth, Western Australia.[128] Instead of using power-hungry cooling towers, the supercomputer is cooled by pumping water from a shallow aquifer, and the pumps themselves run on solar power. Microsoft is building data centres with zero water evaporation via a closed-loop system that recycles water once filled, requiring no new fresh water once the centre is up and running.[129]

When it comes to training AI models, there are active efforts to bring the size, and hence energy demands, of AI models down. Smaller models are more efficient and hence cheaper. Smaller models tend not to perform as well on user tasks, but not every task requires high accuracy and in-depth analysis. We may begin to see a trend where users – or the AI itself – choose the right model for the right job. With this approach, smaller more efficient models can be used most of the time and larger, higher fidelity models are only used when they are really needed.

AI AS A NET POSITIVE FOR THE ENVIRONMENT

So far, this chapter has focused on the negative impacts of AI on the environment. There is another perspective, however. Teams of AI experts around the world are working hard to harness the power of AI to have a positive impact on the environment – to reduce greenhouse gas emissions, adapt to climate change and conserve fauna and flora. Table 13.1 gives just a few of the hundreds of examples available.

AI weather models can now beat the best traditional forecasts.[130] Machine learning can crunch vast amounts of historical weather data to provide more accurate predictions of future weather events, at least in some cases. The current state-of-the-art is that predicting short-term weather patterns based on historical data is feasible. However, predicting long-term climate changes remains an open problem because this isn't simply a matter of looking at historical patterns and projecting them into the future. It is feasible that as AI gets better and is combined with traditional physics-based climate modelling, we will be able to get more accurate and reliable forecasts of the impacts of development on our planet.[131]

AI is also an effective tool in helping society to adapt to climate change. Climate change is predicted to bring more extreme weather events.[132] In places like California, Australia and parts of Europe, this manifests as an increase in frequency and severity of wildfires. An AI tool called Spark is already being deployed in Australia to provide better predictions of the path of a bushfire in real time so that scarce resources can be more efficiently deployed in firefighting.[133]

Human activities can also threaten some of the world's most delicate natural wonders. For example, on the Great Barrier Reef, we are seeing more frequent outbreaks of crown-of-thorns starfish (COTS) that poison and kill coral.[134] A team from CSIRO and Google have come up with an underwater glider, equipped with an AI computer vision system, that can automatically traverse the reef and identify COTS outbreaks, leading to faster removal and limiting reef damage.[135]

When it comes to conservation, AI tends to work best when combined with local expert knowledge. In Kakadu National Park in far north Australia, AI was combined with local Indigenous ranger knowledge to improve the control of a weed grass that was negatively affecting populations of magpie geese.[136] Indigenous knowledge goes back over 65,000 years. By combining it with modern technologies like drones and computer vision, we can get the best of both worlds: deep human expertise working in concert with fast, efficient data analysis.

One of the strengths of AI is its ability to optimise complex processes. Researchers are actively exploring this ability of AI to make power grids more efficient.[137] Modern power grids are incredibly complex. As more renewable energy sources come online, this adds to the complexity because it diversifies where energy comes from but also includes sources that are intermittent. AI has the potential to improve forecasting energy demand, optimise daily planning, and predict outages. Given the complexity of power grids, however, it is unlikely that we will see full automation any time soon; rather, AI will be combined with human operators who can apply decades of experience to tricky edge cases.

There are countless examples where AI is already impacting the environment positively. Whether AI is a net positive or negative contributor remains unclear. As with everything in AI, understanding the different ways that AI can both harm and help the environment is a complex task. Whilst this dichotomy plays out at a societal level, it also arises within organisations, which must consider the pros and cons of AI when it comes to environmental impact and make reasonable decisions about how to use the technology.

WHAT SHOULD YOU DO IN YOUR ORGANISATION

Socially responsible organisations should try to understand the carbon footprint of their activities, including when they use AI. As we've seen in this chapter, to understand the impacts of AI fully it is important to consider the entire supply chain, not just

Table 13.1. Examples of positive contributions of AI to the environment.

Category	**Example**
Climate adaptation	AI-based flood forecasting systems.
Data centres	Reduced data centre cooling by AI optimisation.
Renewable energy management	Improved forecasts of wind farm output.
Wildlife conservation	AI tracking of wild animals.
Sustainable agriculture	Precision weed control to reduce use of pesticides.
Waste management	Robotics and AI sort recyclables with high accuracy.
Pollution monitoring	AI provides real-time air quality data.
Forest restoration	AI-based global forest watch to identify reforestation areas.
Ocean conservation	AI for tracking and removing plastic waste.
Carbon capture	Carbon engineering using AI optimises direct air capture technology.
Disaster response	AI analyses satellite imagery for disaster recovery.
Biodiversity protection	AI-powered sensors detect illegal logging in rainforests.

the end-user task. This means that quantifying the impact of AI is a complex task, but it is important to put some thought into this so you can be sure if your use of AI is a net positive or negative on the environment. The next chapter gives you some pointers on how best to manage this complexity.

QUESTIONS

- In your industry sector do you consider AI to be a net positive or net negative for the environment?
- Consider how your organisation is using AI. Can you identify any significant negative consequences for the environment?

14

How does AI relate to ESG?

ESG (environmental, social, governance) is a term used to describe an organisation's environmental and social impacts. ESG frameworks provide standardised ways to measure these impacts. It has become important to understand ESG for two reasons. First, investors and customers take note of ESG measures when deciding where to do business. Responsible investors want to invest in responsible companies. Second, many countries now have mandatory ESG reporting requirements. For example, in the European Union, the Corporate Sustainability Reporting Directive (CSRD)[138] mandates detailed ESG disclosures for large companies. As a result, thinking about ESG is no longer a nice-to-have but is an essential part of doing business.

ESG frameworks guide companies in understanding and reporting on their environmental, social and governance impacts. Each component of ESG is typically divided into key categories with proposed metrics. For instance, environmental metrics focus on greenhouse gas emissions, resource efficiency and ecosystem impacts. Social metrics cover an organisation's impact on society, such as human rights, workforce displacement, and health and safety. Governance metrics measure how well an organisation manages itself. For example, does it have appropriate policies and reporting in place?

AI adoption affects all aspects of ESG – whether positively or negatively; for example, AI data centres that don't run on renewable energy increase greenhouse gas emissions. On the other hand, AI may be used to optimise logistics, reducing emissions from transportation. Similarly, for social impacts, AI may be net positive or negative, depending on how it is applied. A benefit of using AI would be to provide more efficient customer service for local communities. But if AI is used to displace low-skilled workers without providing alternative job opportunities, the social downsides can be significant. Figure 14.1 gives some examples of how AI can both positively and negatively impact various ESG categories.

Any AI application will have both pros and cons when it comes to ESG. The key is to understand if the pros outweigh the cons. An effective ESG report can really help here. It can provide detailed insights into how a business measures its performance in ESG areas, along with the specific actions it is taking to mitigate

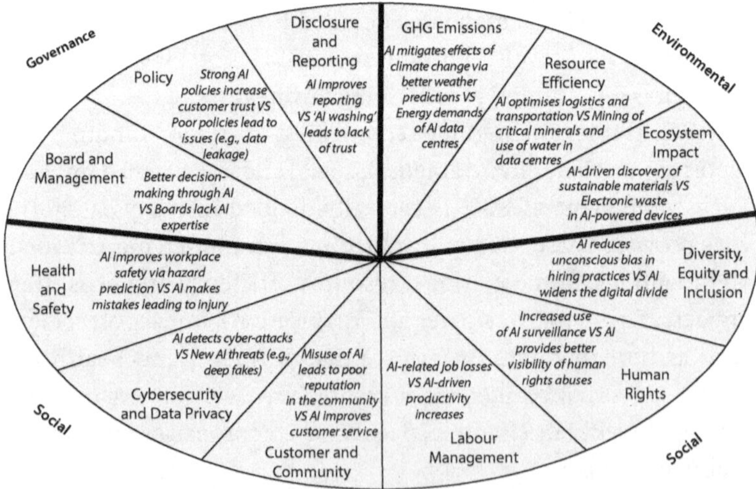

Fig. 14.1. ESG categories and examples of AI positive and negative impacts (adapted from Alphinity Investment Management and CSIRO 2024[139]).

negative impacts and enhance positive contributions. This structured approach helps stakeholders assess a company's commitment to sustainability and responsible practices.

RELATIONSHIP TO EARLIER CHAPTERS

Earlier chapters in this book have covered topics such as AI governance, AI ethics and the environmental impacts of AI. Clearly there is a strong correlation between these topics and ESG. Environmental ESG categories were largely covered in the chapter on AI's environmental footprint. The ESG social impact categories, whilst broad, were discussed in the AI ethics chapter, and the ESG governance categories appear in the AI governance chapter.

What then is the value of dedicating a new chapter to ESG?

Earlier chapters provided general guidance on AI-related social and environmental matters. This chapter, on the other hand, discusses how AI affects your ESG reporting. The key difference between using an ESG framework versus having a general understanding of social and environmental impacts is that ESG focuses on *measuring* and *reporting* impacts so there is transparency for organisational stakeholders.

ESG frameworks use a certain vocabulary and approach for measuring impacts, and this vocabulary can differ, depending on which ESG framework you use. Unfortunately, the language of ESG is different from, say, the language of AI ethics. ESG revolves around a set of concepts (Fig. 14.1) which are related, but don't necessarily map directly on to the concepts from AI ethics literature.

ESG FRAMEWORKS AND REPORTING

Numerous well-established ESG frameworks exist which provide standardised ways for measuring and reporting how your organisation impacts a range of social and environmental aspects.

There is no single preferred framework to use. The choice will depend on your organisation's industry, ESG maturity, expectations of stakeholders and your regulatory environment. Some widely known frameworks are the Global Reporting Initiative (GRI),[140] the Sustainability Accounting Standards Board (SASB)[141] and the Workforce Disclosure Initiative (WDI).[142]

Whilst the details of each of these frameworks differ, the core concepts are the same. Figure 14.2 outlines the basic steps in ESG reporting, no matter which framework you use. The basic idea is to collect data on key ESG metrics that are material for your organisation, assess the data against industry benchmarks to identify areas for improvement, set strategic goals including targets based on the priority metrics and prepare a publicly available report. Many companies include an additional step of external verification, which provides a way that customers and investors can trust the methodology and data in the report.

The main output of any ESG process is an ESG report. A good report should be comprehensive, transparent and relevant, focusing on material issues, using accurate data, and aligning with

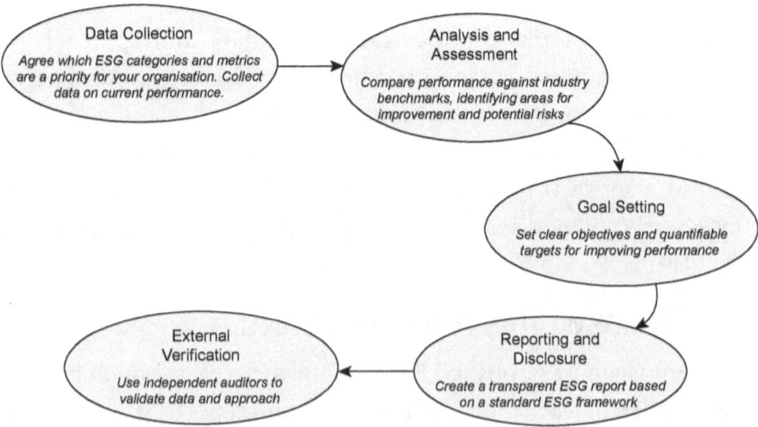

Fig. 14.2. Five stages in development of an ESG report.

industry standards and frameworks. It should also demonstrate how your organisation addresses ESG risks and opportunities.

HOW TO INCLUDE AI IN ESG REPORTING

When adopting AI, you may need to revise your ESG report. Whether you do this depends on the material nature of your AI use cases – are they a significant part of your business that significantly affect your ESG goals or are they relatively minor? As an example, limited adoption of off-the-shelf generative AI tools may not be material and so may not require you to reconsider your ESG report. On the other hand, an organisation-wide AI transformation that affects multiple parts of your business almost certainly will result in the need to revise your ESG report.

The process for reviewing your ESG performance to consider your AI use cases is relatively straightforward. For each AI use case or set of related use cases, revisit the steps in Fig. 14.2 as follows.

1. *Data collection.* Is your set of ESG metrics sufficient now that your organisation is adopting AI? For metrics already in your report, do you need to collect new data? Do you need to consider new metrics not in your current ESG reporting framework? As an example, you may decide to use AI in surveillance applications. This will require you to consider human rights metrics that may not be present in your current reporting. As another example, an AI use case that requires heavy-duty computation in a data centre will require you to recalculate the values of environmental metrics such as greenhouse gas emissions.
2. *Analysis and assessment.* This step remains largely unchanged from the description in Fig. 14.2. You will have collected data on new and/or existing metrics and now should assess your performance against industry benchmarks. You will want to

compare apples with apples here. So, for example, comparing greenhouse gas emission data of an AI-intensive company with an organisation in your industry that doesn't use AI at all probably doesn't make sense. So you may need to re-evaluate which benchmarks you compare against.

3. *Goal setting.* Consider if you need to amend your ESG goals in the light of your AI use cases. This might mean reviewing targets in your goals – some targets may no longer be realistic if you are using AI or may take longer to achieve. Or you may add entirely new goals designed to minimise the ESG impact of AI.
4. *Reporting and disclosure.* The only AI-related considerations here are if there are any voluntary or mandatory AI reporting requirements in your jurisdiction. If there are country-specific laws or guidelines, you will need to decide how to report against these.
5. *Verification and audit.* The process of external validation of an ESG report remains essentially the same. Just bear in mind that not all auditors will have the experience of working with AI, so you may need to change your auditor to one that knows how to quantify the ESG AI impacts.

The essence of this process is to re-examine your ESG reporting framework, understand if AI affects any of the metrics

Top tip: How to incorporate AI into ESG reporting

CSIRO and Alphinity Investments have developed a detailed framework for incorporating AI into ESG reporting.[140] This includes a detailed template for assessing your AI use cases against ESG metrics and a set of over 50 questions designed to prompt your thinking on how AI affects ESG. It is the definitive guide for AI and ESG.

Table 14.1. Re-evaluating ESG metrics against AI use cases.

ESG category	Example metrics	AI use cases positively impacting metrics	AI use cases negatively impacting metrics
Environmental	Energy consumption per unit of output (kWh/unit) % of waste recycled Water usage (litres per employee)	AI-driven energy optimisation in buildings and factories reduces emissions and energy use Predictive maintenance reduces material waste AI supports real-time water leak detection and conservation	AI training (e.g. large language models) consumes high electricity Data centres contribute to carbon footprint Over-reliance on AI-generated environmental models may introduce bias or blind spots
Social	Gender and ethnicity representation in workforce (% of total staff) Employee turnover rate (%) Lost-time injury frequency rate (LTIFR) Number of ethical complaints or violations reported	AI tools help identify bias in hiring and improve diversity metrics AI-enabled safety monitoring reduces workplace incidents Natural language processing (NLP) tools detect early signs of employee dissatisfaction	Algorithmic bias may reinforce existing inequalities Worker monitoring tools may invade privacy or reduce morale Automation may increase layoffs, worsening turnover rates
Governance	Board diversity (% female/ under-represented minorities) Number of confirmed compliance breaches per year Average time to detect/respond to fraud (days) Percentage of policies reviewed and updated annually (%)	AI flags irregular transactions faster, reducing fraud detection time Automated compliance systems improve audit readiness Data analysis supports evidence-based policy decisions	Poorly explained AI decisions undermine board transparency Risk of manipulation through adversarial attacks or flawed data Over-dependence on AI can reduce human oversight in governance

or goals, or requires new goals and metrics, and adjust your due diligence process accordingly. Table 14.1 gives examples of potential AI impacts on ESG metrics. It is important to understand that not all impacts will be negative. Sometimes AI can improve your ESG posture as it brings efficiencies, optimises use of resources and potentially creates new social opportunities.

QUESTIONS

- Does your organisation already produce an ESG report? If so, read through it and mark any places where adoption of AI would require revisions to the reporting framework.
- What reporting requirements is your industry subject to? Think of at least three ways that AI could improve your reporting, either by providing data insights or automating parts of the reporting process.

15

What are the cybersecurity implications of AI?

CYBERSECURITY IS ONE OF THE main issues that keeps board members awake at night.

The number of cyber incidents continues to rise, with a cyber attack now occurring every eight minutes.[143] The scale and impact of these attacks are also escalating. In one case, a breach in a French health insurance system compromised the personal data of over half the country's population.[144] Technology provider Johnson Controls faced a ransomware demand of US$51 million to decrypt stolen data.

Cyber incidents are not limited to targeting consumers; they are increasingly being used as geopolitical weapons. The Centre for Strategic and International Studies (CSIS) maintains a timeline of major cyber incidents since 2006,[145] which makes for sobering reading. It includes not only attacks on corporations but also breaches affecting governments, critical national infrastructure and even the core institutions of democracy.

The upshot is that any company should not be asking itself *if* it will be attacked, but *when*. And if that statement sounds scary, wait until we bring AI into the equation.

AI fundamentally changes the cybersecurity profile of your organisation. Malicious actors are using AI to carry out cyber attacks faster and at greater scale. AI's ability to automate routine tasks means that attacks can now be scaled up for a fraction of the cost. AI also opens up new vulnerabilities in IT systems. Given its statistical nature, AI can be unpredictable, which means unexpected gaps in your cyber defences. If you can't explain exactly how an AI system works, how do you know if it's secure?

AI isn't all bad news for cybersecurity though. AI has already become a key weapon in the fight against cybercrime. Cyber-defence tools now leverage AI, providing better protection against attacks.

As is often the case with AI, its relationship with cybersecurity is complex. On the one hand, AI can be a powerful force for good, enhancing your organisation's ability to detect, prevent and respond to threats in an increasingly volatile cyber landscape. On the other hand, it introduces a new set of risks. As a leader, you'll need to understand these emerging threats and ensure your cyber incident response plans are updated to include AI-specific mitigations.

AI AS A CYBER DEFENDER

The cybersecurity industry is one of the strongest adopters of AI. It's easy to see why. Thwarting cyber attacks is a difficult business. The United States alone has ~550 exabytes (about half the number of grains of sand on the planet) of internet traffic every month, and the rate of growth is exponential. Every single internet interaction is potentially vulnerable to an attack. It is clearly impossible for humans to monitor all that traffic so this is the perfect application for AI.

AI is used in just about every aspect of cybersecurity. It is used to detect abnormal patterns in internet traffic that might be a sign of an attack. It is used to automate responses to a cyber incident – AI can respond much faster than your IT department. AI is also used to predict future attacks as an AI crystal ball for cybersecurity.

Beyond direct cyber attacks, AI is widely used in detecting fraud. It has also become a key technology for access control, for example, through biometric identification. In fact, it is impossible for me to cover all the ways that AI is being used in the cybersecurity industry in this book – that would require a book in itself.

Without attempting to cover all the benefits of AI in cybersecurity, I will describe just a few of the more interesting examples.

There are four key phases that organisations must consider in cyber preparedness: understand vulnerabilities, detect incidents, recover from incidents and learn for next time. AI can be a useful tool in any of these phases.

Understand vulnerabilities

Preparing for a cyber incident requires an organisation to understand its vulnerabilities fully. It is only by understanding potential threats that an organisation can decide on the best responses. All organisations are vulnerable, but the specifics depend on your context. For example, an online retailer dependent on website availability for sales is vulnerable to denial-of-service attacks. In contrast, a mining company may be more worried about specialist IT systems at mines.

Typically, understanding vulnerabilities relies on a systematic process to enumerate potential threats and then assess if the organisation is vulnerable to those threats. This is a highly intensive process, requiring the knowledge of experts from different fields. It is a great application for AI. AI has access to vast datasets of information about how cyber attacks happen. By using AI, companies can produce more thorough and reliable models of likely threats.

AI is not a replacement for human expertise, though. AI is best seen as a complement to human expertise. For example, generative AI tools like ChatGPT can be used as cybersecurity advisors to enumerate possible threats to an organisation quickly. The more information you give to ChatGPT about your business, the better it will be at suggesting potential threats. While generic tools like ChatGPT can be a good aid if used properly, there are also generative AI tools that have been specifically designed for modelling threats.[146]

Detect incidents

A cyber incident response plan should outline steps your organisation is taking to detect cyber incidents. Typically, threats identified in the previous phase will be captured and consideration will then be given on how to detect those threats in practice. This involves monitoring activity and alerting anything suspicious.

A typical use case for AI in detecting incidents is *anomaly detection*. Anomaly detection is where AI is used to detect an unusual pattern in a stream of data. The most familiar example of anomaly detection is in credit card fraud. We have all received automated calls from our bank when there's unusual activity on our card. Credit card companies now routinely use AI both to detect unusual patterns but also reduce the number of false alarms that might annoy customers.

More generally, AI-based anomaly detection is used to identify unusual patterns in network traffic that may be signals of an impending cyber attack. And most email systems nowadays use AI to filter out spam and flag possible cybersecurity issues, such as phishing attempts.

Recover from incidents

The objective of this phase is to limit the damage caused by a cyber incident, remove the threat and restore normal operations. Again, AI can be a useful tool because automation means that operations can be restored much faster than relying on human intervention.

There are many ways that AI can be used to automate the containment or recovery from an attack. A somewhat sophisticated but interesting approach is to use AI-driven 'honeypots'. In cybersecurity, a honeypot is a decoy system set up to entice cyber attackers so their behaviour can be monitored and prevention measures put in place. For example, an organisation might set up a fake database with imaginary customer details and wait for a malicious actor to attempt to gain access to the database.

Honeypots have been used as a cybersecurity defence mechanism for a long time. We are beginning to see the emergence of AI-driven honeypots. For example, AI might be used to improve simulation of a real system so that attackers don't realise the simulation is a fake. Or AI might be used to learn from an attacker's behaviour and suggest cyber prevention techniques.

Learn for next time

A critical aspect of any cyber incident response plan is to learn from mistakes. If incidents occur, it's important to document these to ensure the organisation is better prepared in the future. AI's contribution here could be looking for patterns in incident data to identify areas that need most attention. Most organisations are subject to tens, if not hundreds, of cyber attacks every month. This can result in a lot of data that needs to be understood to implement better prevention mechanisms. AI is well suited to this task.

As you can see, AI is becoming a core weapon in the defence against cyber attacks. In fact, traditional AI has been used in cybersecurity for decades. More recently, generative AI is bringing new ways of preventing cyber incidents. For example, large language models can learn the characteristics of insecure computer code. These LLMs can then be used to automate security testing of a company's software so that it contains fewer vulnerabilities.

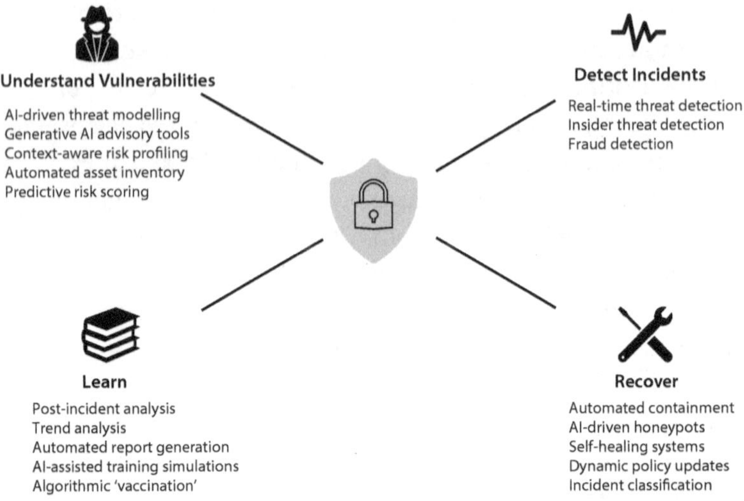

Fig. 15.1. Ways AI is defending against cyber attacks.

I will end this section with one particularly interesting example where AI is being used in cybersecurity, which I call vaccines for algorithms.

An *adversarial attack* occurs when a malicious actor deliberately targets the AI system itself. In the case of machine learning, this often involves manipulating input data in subtle ways to deceive the system and trigger unwanted behaviours. For example, imagine a secure access door in a factory that uses AI-driven voice recognition. An attacker could embed a barely audible audio signal into a voice recording – inaudible to humans, but enough to trick the AI into misclassifying the intruder's voice as authorised.

To defend against such attacks, researchers are developing new techniques that effectively 'vaccinate' AI systems.[147] These methods involve training the AI on slightly corrupted or adversarial data so that it learns to recognise and resist similar attacks in the real world. Much like a biological vaccine, this approach helps the AI build resilience against future threats by being exposed to harmless versions of them during training.

Figure 15.1 summarises the many ways in which AI is being used in the defence against cyber attacks.

AI AS A CYBER ATTACKER

While AI is a new weapon in our cyber defence capabilities, it also brings fresh cybersecurity challenges. There are two aspects to this. First, AI makes it easier and more efficient for bad actors to carry out attacks at scale. Deepfakes are an example. Almost anyone can now generate images or even video of fake news coverage that can be used to mislead the public. Second, given the statistical nature of AI, these systems may be unintentionally vulnerable to attack. No one understands the exact inner workings of an AI system. As a result, AI can sometimes produce unexpected or unusual results, and this fact can be exploited by attackers to coerce an AI into problematic behaviour. As an example, in 2024, Google rolled out its Gemini generative AI tool across the Google Suite, making it available on all its office products, including Gmail and Google Docs. Within 30 minutes of the launch, a team of IT professionals 'hacked' Gemini, making it generate misinformation in emails and include phishing links in corporate communications. Despite years of work in putting guardrails into generative AI tools to avoid such behaviour, the problem remains.

Deepfakes

Deepfakes are synthetically generated images, audio or video, that are used to promote misinformation and cybercrime. Early deepfake technology was relatively easy to detect. Subtle clues in the generated content were a giveaway that it was AI generated – for example, inconsistencies in lighting, misalignment of spoken words and lip movements, or poorly rendered body parts. However, modern deepfake technology is so good that it is now almost impossible to distinguish between real and fake. There is a growing market in deepfake detectors (themselves based on AI technology), which can be effective in some applications, but the average person is unlikely to be able to spot a deepfake consistently.

Deepfakes can have a positive side. They have been used in entertainment and education – for example, to bring to life historical figures and teach history in a more interesting way. A UK TV program, *Deep Fake Neighbour Wars*, used AI to superimpose celebrity faces on video of real actors, thus creating comedy scenes in which actors like Kim Kardashian and Idris Elba find themselves in unusual situations.[148]

Most applications of deepfake technology, however, are malicious. 96 per cent of deepfakes are used in pornography, leading to a new type of gender discrimination.[149] The use of deepfakes in spreading political misinformation and in carrying out fraud is increasing. A key risk is the use of deepfakes to fool digital authentication systems such as voice or facial recognition systems used to secure online assets. A deepfake image of your face can be used to gain access to your mobile phone. Banking systems that rely on voice authentication are vulnerable to deepfake voice clones.

Perhaps the scariest thing about deepfakes is how easy they are to produce. In 2023, a mother in Arizona got a call from a kidnapper, claiming he had her teenage daughter and demanding a ransom. The kidnapper put the daughter on the phone who, sobbing, begged her mother for help. As it turned out, the daughter was in her father's house, safe and sound. The 'kidnapper' used a deepfake voice clone, using audio from the daughter's YouTube channel to train an AI system.[150] And this is not an isolated incident. A recent UK study found that 28 per cent of adults have faced voice cloning scams.[151] Anyone with a YouTube or TikTok account is potentially at risk. AI needs as little as three seconds of audio to generate a high-quality deepfake.[152]

Speed and scale

The real game-changer that AI brings is that it vastly increases the speed and scale at which bad actors can carry out attacks. Most of us have been subjected to scam phone calls from someone trying to convince us to give away personal details. (When I receive those calls, I find it quite amusing to play along and see how long I can keep the scammer on the phone before they hang up on me.) Traditionally, these scams required real people in call centres somewhere across the world making hundreds of phone calls in the hope that one or two vulnerable people would be fooled. But with AI you don't need real people. Fake voices can hold a conversation with you in real-time and can often be more convincing than a human scammer because the AI can be more persuasive. A CSIRO study has shown that AI can be very good at manipulating human decision-making.[153] Generative AI is being used to generate fake emails, text messages and voice messages on a large scale. The life of an average cyber scammer has become a lot easier.

One of the 'killer apps' of AI is its ability to generate computer programs. This capability is transforming the software industry. Amazon has claimed that AI saved it 4,500 years in programming time.[154] But if AI can be used to generate code for the software industry, it can also be used by bad actors to generate malicious code faster than ever before, thus putting IT systems' security at risk. In 2024, Microsoft found evidence of foreign governments using generative AI to create malware and phishing campaigns.[155] In fact, publicly available generative AI tools are routinely used to generated computer viruses and malware. Even though most tools have safety measures in place to avoid this, they are not difficult to circumvent.

Unintended vulnerabilities

AI is often described as a 'black box' technology, meaning that it cannot explain in terms that are easy to understand all the steps that lead to a particular output. This 'black box' nature stems from the inherent internal complexity of large-scale AI systems, meaning it can be impossible to predict exactly how an AI system will behave once deployed. This means AI is hard to test. And it is hard to know if something will go wrong once the system is deployed. As a result, it is quite likely that an AI system will exhibit behaviours that can be exploited by an attacker.

Take AI-generated software code as an example. Several studies have shown that AI-generated computer programs contain errors, sometimes in very high numbers. For example, code generated by GPT-4 can contain up to 60 per cent misuses of common coding frameworks.[156] Errors in generated code aren't necessarily a problem if human developers thoroughly check and test the code. However, when it comes to cybersecurity, security flaws in code, which make computer systems easier to hack, are often subtle and require a sophisticated testing approach to uncover. This means that AI-generated security flaws can easily be missed, and we could be unleashing thousands of unsecure computer programs into the world without even knowing it.

These unintended vulnerabilities are not limited to code. Any generative AI system can result in vulnerabilities when uninformed users unwittingly do something that causes security or privacy issues. A common concern of organisational leaders is when employees enter private or sensitive company data into generative AI tools. Many such tools retain user data to train their AI models. This means that data can leak outside the company and be regurgitated in someone else's chatbot response. In 2023, Samsung employees accidentally leaked confidential information when using ChatGPT to review internal documents.

Users who enter personal information into AI chatbots have sometimes found their information disclosed to other users. A study showed that almost 10 per cent of generative AI prompts contain sensitive information, such as customer data, finance data and sensitive code.[157] Although there are ways of using AI tools in 'incognito' mode where user prompts are not stored or used for further training, and enterprise-grade AI tools have security features built in to avoid data leakage, around 60 per cent of users use free or low-cost versions of AI tools without these safeguards.[158] InfoSecurity magazine has reported that nearly a fifth of CSIOs said their staff had leaked data using generative AI tools.[158]

The only way around data leakage by employees is to introduce policies on corporate use of AI tools. A minority of organisations currently have such policies,[159] meaning that 'shadow' AI use, whereby staff are using AI without anyone knowing, is commonplace.

WHAT YOU SHOULD DO TO AVOID AI CYBER INCIDENTS

This chapter has described how AI can be both a source of cybersecurity issues but also a guard against them. This leaves you, as a leader, in a quandary. Do you embrace AI in your organisation, hoping the business benefits of AI will outweigh any cybersecurity risks, or do you take a cautious approach, risking missing out on the opportunities that AI may bring?

The answer, of course, is to adopt AI where it brings tangible business benefits, and to do so in a way that takes full account of any cybersecurity risks.

So how can you do this as a leader?

The only way to be prepared for AI-related cyber incidents is to review and update your cybersecurity policies and procedures accordingly. This requires carrying out the following steps:

1. Review cybersecurity policies to ensure that AI-related incidents are covered.
2. Review your cyber incident response plan to ensure that AI-related incidents are covered.
3. Introduce policies on the use of AI within the organisation.
4. Update your risk register to capture any changes from (1), (2) and (3).
5. Communicate changes to staff and provide training where necessary.

The range of potential cybersecurity risks from AI is wide. Table 15.1 captures common examples of these risks. My suggestion is that you take a systematic approach to review your cybersecurity procedures for each of these AI-related risks. Do the risks in Table 15.1 apply to your organisation? If so, what mitigations will you put in place? Are there AI cybersecurity tools you can use to protect your organisation? And how will you convince your management and/or board that you have fully considered both the cybersecurity benefits and risks of AI?

I wish you the best of luck. Remember: AI can be a great boon for your business, but only if governed properly, and that governance should absolutely include a rigorous analysis of the cyber risks.

QUESTIONS

- What worries you most, from a cybersecurity perspective, about adopting AI in your organisation?
- What mitigations could you put in place to alleviate these worries?

Table 15.1. Potential cyber risks from adopting AI.

Risk category	Risk	How AI adoption affects risk profile
Human-related risks	Phishing and social engineering	AI-powered customer support chatbots might be impersonated to steal customer credentials.
	Insider threats	AI monitoring employee behaviour could lead to misuse if an insider alters configurations to target colleagues.
	Password weaknesses	AI-based facial recognition for access control could be bypassed using deepfake videos.
	Untrained employees	Employees relying on AI-generated reports might misinterpret results, leading to security oversights.
Network risks	Man-in-the-middle (MITM) attacks	AI-enhanced firewalls may misclassify genuine traffic as safe, exposing sensitive data.
	Distributed denial of service (DDoS)	AI-driven customer analytics platforms hosted online could be taken offline, disrupting services.
	Unauthorised access	AI-based automated systems may misidentify attackers as legitimate users.
	Unsecured Wi-Fi networks	AI-controlled Internet of Things (IoT) sensors on a Wi-Fi network may be exploited to leak operational data.
Endpoint risks	Malware and ransomware	AI-driven employee productivity apps may inadvertently execute ransomware if compromised.
	Outdated software	AI in legacy manufacturing equipment may be exploited to cause operational delays.
	Lost or stolen devices	AI-enabled mobile devices with sensitive client data may be exploited if lost or stolen.
	Unsecured IoT devices	AI-managed smart devices in offices could be accessed to disrupt physical environments.

continued

Table 15.1. continued

Risk category	Risk	How AI adoption affects risk profile
Application risks	Web application attacks	AI-driven customer portals could expose personal data if a vulnerability is exploited.
	API vulnerabilities	AI systems integrating with third-party APIs may inadvertently expose sensitive transaction data.
	Supply chain risks	AI software automating procurement decisions could be manipulated to favour malicious vendors.
	Shadow IT	Employees may use unauthorised AI tools (e.g. ChatGPT) to process confidential information.
Data risks	Data breaches	AI-powered customer relationship tools might store sensitive data that attackers can target.
	Data exfiltration	AI systems generating reports might inadvertently send data to unauthorised email addresses.
	Unencrypted data	AI systems analysing customer behaviour might store sensitive data in unencrypted formats for faster processing.
	Data loss	AI tools performing automated backups might overwrite critical data during system errors.

Risk category	Risk	How AI adoption affects risk profile
Cloud-specific risks	Misconfigured cloud services	AI workloads deployed on misconfigured cloud platforms could expose training datasets.
	Shared responsibility misunderstanding	Organisations might assume cloud providers secure AI models, leading to gaps in protection.
	Account takeovers	AI credentials stored in shared cloud services could be compromised, granting attackers access to models and data.
	Data residency issues	AI analysing data across regions might violate local data privacy laws, such as the EU GDPR.
Supply chain and vendor risks	Third-party breaches	AI systems relying on external data providers may propagate breaches into internal systems.
	Dependency risks	AI software depending on single-vendor APIs might fail if the vendor is compromised or goes offline.
	Hardware backdoors	AI accelerators used in critical systems might contain hardware backdoors, exposing operations.
Regulatory and compliance risks	Non-compliance penalties	AI systems generating customer data insights might fail to anonymise data, breaching regulatory requirements.
	Legal risks	AI used in financial modelling could produce erroneous reports, leading to lawsuits from stakeholders.
	Reputation damage	Staff use AI to generate press releases, which may contain errors, leading to public backlash and loss of trust.

continued

Table 15.1. continued

Risk category	Risk	How AI adoption affects risk profile
Advanced threats	Advanced persistent threats (APTs)	AI managing critical infrastructure could become a target for prolonged, undetected attacks.
	Zero-day vulnerabilities	AI integrated with IoT devices could expose unknown vulnerabilities that attackers exploit.
	Cryptojacking	AI workloads in cloud environments might be hijacked for cryptocurrency mining without detection.
	Deepfake attacks	AI-enabled identity verification could be tricked by deepfakes to grant unauthorised access.
Physical and environmental risks	Physical theft	AI-powered delivery robots might be stolen to extract sensitive data stored onboard.
	Natural disasters	AI systems managing disaster response might misinterpret sensor data, leading to delayed action.
	Power failures	AI systems controlling manufacturing might fail during power outages, halting production.
Strategic risks	Reputational damage	AI chatbots managing customer support might respond inappropriately, causing reputational harm.
	Financial risks	AI used in automated trading could make high-risk trades during market anomalies, resulting in significant losses.
	Operational downtime	AI systems optimising production schedules might miscalculate, leading to costly downtime.

Part V

Bringing it all together

To wrap things up, this final part of the book uses a case study to illustrate some of the key concepts when adopting AI in your organisation. The first chapter looks at a scenario where AI adoption is done correctly and well. The second chapter examines how things can go wrong. In the appendix, you can find a detailed analysis showing how each chapter in this book can be applied in a practical way to the case study.

Note: these two case study chapters were written by ChatGPT. Here is the conversation with ChatGPT 4o which led to these chapters.

> Jon: I have written a book. I want to add a final chapter that illustrates key concepts in the book

with a case study. If I upload the book can you write the final chapter for me?

ChatGPT: Absolutely – if you upload the book, I can read through it and draft a compelling final chapter based on a case study that ties together the key themes and concepts.
A few things that will help me tailor it well:

1. *Your target audience* – Is it business leaders, general readers, technologists, policy-makers?
2. *Tone and style* – Formal? Conversational? Inspirational?
3. *Preferred case study format* – Real-world company? Fictional but plausible scenario? A blend?
4. *Any specific concept(s)* you definitely want highlighted?

Once you upload the manuscript, I'll take it from there.

Jon: [uploads book draft]

ChatGPT: Great – thanks for uploading the manuscript. Based on what I've seen, your book is practical, insightful and grounded in real-world leadership experience with AI. A final

chapter that illustrates key concepts through a case study will be the perfect way to bring it all together.

Before I draft it, just a few quick questions to tailor it exactly to your needs:

1. **Would you prefer a real company (e.g. Telstra, BCG, or Air Canada) or a fictional but plausible case?**
2. **Do you want the tone to match the rest of the book – conversational, with light humour and top tips?**
3. **Should the case study showcase:**
   - **the entire journey from early interest to adoption at a large scale?**

     or
   - **focus on a particular theme (e.g. AI governance, productivity, customer value)?**
4. **Would you like the final chapter to include a template or framework (like a one-page AI adoption canvas) that readers can use?**

Once you confirm, I'll write a complete chapter draft in your voice.

Jon: 1. use CareHouse as the name of a fictitious company; 2. yes; 3. entire journey; 4. yes please!

[ChatGPT writes first chapter]

Jon: now give me a version of this same story where things go wrong.

[ChatGPT writes second chapter]

The case study is followed by a wrap-up section for the book, summarising key takeaways for each chapter. This was written with the assistance of ChatGPT to summarise chapter content but with heavy human editing of the result.

The appendix was written by me but with ChatGPT's help to generate the tables.

16

A successful case study

LET ME INTRODUCE YOU TO CareHouse.

CareHouse is a fictitious mid-sized healthcare provider based in regional Australia. Its mission is to improve patient outcomes through personalised, human-centred care. It specialises in at-home services, including nursing, physiotherapy and post-operative support. Like many healthcare providers, CareHouse faced an increasingly complex operating environment: rising demand, workforce shortages, tight budgets and changing patient expectations.

The company knew it needed to do something different. And it thought AI might be part of the solution.

This chapter tells the story of CareHouse's AI journey from curiosity to capability, from pilot to practice, and from hype to impact. It isn't a perfect story. The company made mistakes. It adapted. But through it all, the company kept its focus on one thing: delivering better outcomes for their patients.

HOW IT STARTED

CareHouse's interest in AI was first ignited by the CTO, Rohan, who had been experimenting with generative AI tools in his spare time. He used ChatGPT to summarise meeting notes, draft code snippets, and even help brainstorm new service models. One day, he used it to rewrite a patient communication in plain language, and something clicked.

'What if we could scale this up?' he wondered. 'What if every patient received easy-to-understand, personalised care instructions, instantly and accurately?'

Rohan brought his idea to the executive team. There was initial scepticism, particularly from the Chief Risk Officer, but the CEO, Angela, was curious. She asked a simple but powerful question: 'Could AI help us deliver care that's more personal, not less?'

THE EXPLORATION PHASE

CareHouse began in the same way many organisations do: with exploration. The company ran a series of lunch-and-learn sessions for staff, where AI was demystified. A cross-functional working group, with representatives from nursing, IT, legal, operations, and patient advocacy, was set up, and staff were surveyed in order to understand pain points that could be addressed with AI.

Three themes emerged:

- *Information overload.* Staff were overwhelmed by the volume of patient data and protocols.
- *Communication gaps.* Patients often didn't understand post-care instructions.
- *Inefficient scheduling.* Too many missed visits, double bookings and travel inefficiencies.

Each of these became a candidate use case. But rather than jump into building tech, CareHouse took a pause. The company wanted to do it right.

THE WHY

Angela insisted they clarify their 'why'. The team developed a simple values statement:

> 'We will use AI to amplify our human touch, not replace it. Our goal is to make care more personalised, accessible and effective for every patient we serve.'

The company also defined success metrics: improved patient satisfaction scores, reduced nurse admin time, fewer missed visits and better health outcomes.

THE FIRST PILOT

The first experiment was small. CareHouse partnered with NeuroAI, an Australian AI startup, to pilot an AI-driven remote patient monitoring system. The idea was simple: patients would wear discreet sensors at home. The system would analyse movement, sleep and vital signs to detect early signs of deterioration.

The tech was sound but the pilot revealed important lessons:

- The AI flagged too many false positives.
- Nurses worried they were being replaced.
- Patients didn't know how their data were being used.

Rather than press on blindly, CareHouse took a step back. The model was retrained, co-design sessions with staff were run, and a plain-English data consent form was developed.

Patient satisfaction jumped. Staff buy-in improved. And the pilot was extended.

SCALING SMART

Having learned from the pilot, CareHouse began scaling up carefully. The company adopted a three-stream strategy:

1. *Low-risk productivity gains.* CareHouse introduced a generative AI assistant for internal use to help staff draft reports, update patient records and manage rosters.
2. *Customer-facing augmentation.* The company built a chatbot trained on internal documents to answer routine patient queries.
3. *Strategic AI capability.* The organisation began investing in a central AI governance unit and training clinical staff in basic AI literacy.

Importantly, CareHouse didn't treat AI as an IT project; it was treated as a whole-of-business transformation.

GOVERNANCE AND RISK

From the beginning, CareHouse embedded ethics and governance into its AI journey. It created an AI Oversight Panel including clinicians, ethicists, patients and technologists. Every AI application was assessed for fairness, transparency, explainability and alignment with company values.

CareHouse adopted a simple but powerful principle: *If we can't explain it, we won't deploy it.*

This led the company to reject several off-the-shelf tools that lacked explainability, even though these tools promised high performance.

THE RESULTS

Eighteen months into the project, CareHouse had:

- Reduced nurse administration time by 30 per cent
- Increased patient comprehension of discharge instructions by 45 per cent
- Achieved a 22 per cent reduction in hospital readmissions
- Seen staff engagement scores rise for the first time in years

The AI journey was far from over, but CareHouse had made it through the hardest part: starting.

REFLECTIONS FROM THE CEO

Angela later said: 'AI hasn't made us less human. It has made us more human. Our people are spending less time typing and more time caring. That's what this is all about.'

17

A not so successful case study

MEET CAREHOUSE, AGAIN.

This is the same mid-sized healthcare provider you met in the previous chapter. Same mission: to deliver personalised, high-quality care to patients in their homes. Same ambition: to use AI to enhance their services, reduce costs and improve outcomes.

But this time, things didn't go to plan.

THE RUSH TO INNOVATE

It began with a board meeting.

Angela, CareHouse's CEO, had just returned from a health tech conference where everyone was talking about generative AI. She had seen demos of chatbots writing discharge summaries, AI tools automating triage, and predictive analytics promising to reduce readmissions by half.

Fearing they might fall behind, she told her executive team: 'We need to do something with AI. Now.'

The mandate was clear: move fast. But fast meant skipping steps.

NO CLEAR PURPOSE

The first red flag? No one agreed on why they were adopting AI.

The CIO thought it was about cutting costs. The Chief Nurse hoped it would reduce paperwork. The COO saw it as a branding exercise. And the marketing team wanted to announce something flashy.

The company started multiple projects at once:

- A chatbot to handle patient queries
- An AI scheduling tool for home visits
- A natural language processing tool to summarise nurse notes

But there was no overarching strategy. No definition of success. And no alignment on risk appetite.

IGNORING THE HUMANS

The chatbot was built quickly using a third-party vendor. It was trained on internal documents, but without assessing the quality of those documents first. The result? Inaccurate answers. Patients were told to take medications at the wrong times. One patient missed a critical follow-up.

The nurses were furious. 'We're being replaced by a tool that doesn't even work,' one said.

Meanwhile, the scheduling AI, intended to optimise nurse visits, reduced travel time on paper but led to unrealistic rosters. Nurses were booked for back-to-back visits across suburbs. Morale plummeted. Resignations spiked.

No one had involved frontline staff in the design.

THE DATA MESS

CareHouse had lots of data, but it was messy. Records were duplicated, outdated or incomplete. Still, the team forged ahead.

The AI tool that summarised nurse notes generated summaries that missed key clinical details. One nurse wrote in a patient's record: 'Monitor for possible signs of infection due to recent surgery.' The AI summary? 'Patient mostly stable.'

This was only caught because a senior clinician double-checked the notes, but the damage was done. Trust in the system eroded.

ETHICS AS AN AFTERTHOUGHT

There was no ethics review. No privacy impact assessment. No explainability testing. The team assumed that using a reputable vendor was enough.

But a journalist had used patient conversations to train their own models, and some patients could be identified. The story made headlines. Patients were outraged. Regulators started asking questions.

Angela faced the board.

'We wanted to be leaders,' she said, 'but we've lost the trust of our staff and our community.'

WHAT THEY LEARNED

The AI projects were paused. An external review was commissioned, and the results were sobering:

- *Poor planning.* No AI roadmap or defined outcomes.
- *Lack of engagement.* Minimal input from staff and patients.
- *Immature data practices.* Dirty data fed into unreliable systems.
- *Governance gaps.* No oversight, no risk mitigation.

In short, they had skipped the basics.

REBUILDING

CareHouse didn't give up on AI, and it started again, this time with humility.

The company created a cross-functional AI steering group. It embedded co-design into every project. It developed data standards and appointed a Chief Data Officer and established an AI ethics panel.

Angela's new mantra?

'AI is not a race. It's a responsibility.'

FURTHER DETAILS

The CareHouse example shows how AI can benefit an organisation and what can go wrong. The appendix to this book provides a lot more detail on how CareHouse could go through the steps outlined in each section of this book to define, execute and govern a good AI strategy.

18

If you remember only one thing ...

THIS BOOK HAS PROVIDED A practical roadmap to help organisational leaders navigate the complex and rapidly evolving world of AI adoption. We have covered everything from strategy to execution, from governance to culture, from ethics to cybersecurity.

In this final chapter of the book, I reflect on each book section and draw out what I think are the key takeaways. If you remember only one thing from each chapter, let it be the following.

PART I – CONTEXT

Key takeaway 1: AI isn't magic; it's just maths. Understanding the limitations of AI helps you, as a leader, cut through the hype. AI isn't intelligent. Instead, think of AI as large-scale statistical pattern matching.

Key takeaway 2: AI is great at some things and not so great at others. As a statistical engine, AI sometimes makes mistakes. That's fine for some applications but not for all.

Key takeaway 3: How fast is AI progressing? *Very* fast. The technology itself is rapidly evolving but the organisational change needed to leverage AI fully is much slower. We are only now beginning to see how quickly AI adoption will permeate business.

PART II – WHY AI?

Key takeaway 1: AI can boost productivity on individual tasks, but it is more challenging to translate that into organisation-wide productivity improvements. This takes effort, persistence and a whole-of-organisation approach that also considers cultural change and process improvement.

Key takeaway 2: Cost savings from AI are possible but don't forget the investment required to get there. The technology itself is relatively cheap but expenditures of IT infrastructure, training, governance and cultural change can quickly blow up.

Key takeaway 3: One of the best ways to think about AI is how it can add value for your customer. Try to think beyond just automation. What are the major challenges your customers are dealing with and can AI help to solve those?

Key takeaway 4: True disruption happens when AI enables things that weren't previously possible, not just faster versions of what already exists.

PART III – AI STRATEGY

Key takeaway 1: Your AI strategy must be grounded in purpose, aligned with your values and business goals and owned by leadership. Start with the problem you are trying to solve, not where you can easily apply AI.

Key takeaway 2: Structuring your organisation to succeed with AI is key, but there's no one-size-fits-all solution. Choose a structure that matches your AI maturity and your culture.

Key takeaway 3: Governance of AI is critical! Don't think of governance as slowing things down; it's how you scale AI safely and sustainably to your organisation and its goals.

PART IV – THE BIGGER PICTURE

Key takeaway 1: Ethical AI starts with values. Be clear on your values and those of your customers and stakeholders. Use values as your anchor to make decisions on where, when and how to apply AI.

Key takeaway 2: AI can be energy intensive. Don't ignore the carbon cost of AI innovation.

Key takeaway 3: Every AI application has ESG impacts. Your job is to be clear and transparent on these so you can maximise the positive and minimise the negative.

Key takeaway 4: AI can be a cyber defender of your systems, but it can also be attacked. Cybersecurity and AI must evolve together.

FINAL WORDS

Back in 1994, I was 22 years old, fresh out of an undergraduate mathematics degree, sitting in a careers library, wondering what to do with my life. Like most of you, many of the major decisions in my life have been serendipitous. My journey would have been very different had I not flicked through a particular magazine and seen an ad encouraging me to study artificial intelligence.

Even back in 1994, AI was not new, but it was new to me. And it felt fresh, exciting, and something that could change the world. In the 30 years since, I've been fortunate to work on some groundbreaking AI projects that have changed the world for the better.

In the early days of my career, I taught computers how to write code for space missions. I taught them how to solve mathematical equations. I worked with air traffic control experts using AI to automate the movements of airplanes in the sky. Many of those problems were *really* hard at the time, so hard that it felt like they would never be solved. But now they have been.

Later in my career, I took on more of a leadership role. I don't develop my own AI systems anymore but my teams have used AI to help conserve the Great Barrier Reef, to create an AI assistant for astronauts that is currently flying on the International Space Station, to save lives of first responders in search and rescue operations, and to help communities adapt better to a changing climate.

One thread that has run through all this work is the importance of putting the human at the heart of technology. There are plenty of pundits who see AI as a replacement for humans. I don't see it that way. Not because I'm afraid of it but because I believe the real power of AI lies not in replacing us, but in *amplifying* what makes us human: our compassion, our creativity, our judgement. What excites me isn't machines thinking like people but people using machines to think more boldly, act more wisely and connect more deeply.

To achieve this, we must always put people first. That applies whether you're a technologist building the next wave of AI, a policy-maker shaping how AI influences society or a business leader adopting AI to improve operations.

So as a business or organisational leader embarking on an AI transformation, my biggest piece of advice is simply this: always put the human, people, first. Focus on your values and how AI can enhance them. Resist the temptation to compromise those values for a quick win.

If you don't internalise this, everything else I've written in this book is irrelevant.

There have been many points in my career when I was tempted to stray from this path. Lucrative opportunities came along. Projects that promised glory but didn't sit right with my gut. It would have been easy to say yes to many of these. It is much harder to say no.

And that's exactly how you need to approach AI in your organisation. Not with blind excitement, but with clear intent. Be bold enough to imagine what's possible, but brave enough to walk away when it doesn't feel right. Just because AI *can* do something doesn't mean it *should*. You will be told to move fast, to automate everything, to chase the next big thing. But in those moments, pause. Remember what you are really here to do.

AI will change your business. But whether it changes it for better or worse, that's up to you.

I've seen AI navigate Mars and I've seen it fight wildfires. I've seen it solve long-standing scientific problems and I've seen it bring joy to a sick child. But what I've come to value most is not the sophistication of the systems but the conviction of the leaders behind them.

So lead with clarity. Lead with courage. And above all, lead with your values intact.

That's how you make AI work for people. And that's how you build something that lasts.

Appendix

CareHouse is a fictitious healthcare company that focuses on improving patient outcomes through personalised care. The company provides at-home healthcare services, including nursing, physical therapy, and post-operative care. Traditionally, CareHouse has done this through its network of contractor nurses, who spend a lot of their time on home visits. In the previous 12 months, CareHouse has partnered with an AI company, NeuroAI, using AI to monitor patient health remotely through wearable devices that track vitals and symptoms. AI algorithms analyse the data in real-time, alerting healthcare professionals if intervention is needed and adjusting care plans based on the patient's condition.

CAREHOUSE'S VALUES AND THE IMPACT OF AI

Table A.1 summarises CareHouse's values and the potential impacts on those values, both positive and negative, of partnering with NeuroAI. These are important to understand so that CareHouse can make an informed judgement whether to partner with NeuroAI at all. The table also includes some suggested mitigations so the decision-maker can understand if risks from the negative impacts are manageable. This analysis can also be complemented by a risk assessment that quantifies the risk associated with each value from adopting AI (Table A.2). Based on the residual risks, CareHouse can decide whether to proceed with the NeuroAI partnership.

Table A.1. The impact of AI adoption on CareHouse's values.

Core value	Description	Potential positive AI impacts	Potential negative AI impacts	Mitigations
Patient dignity	Valuing patients' dignity in personalised care plans.	AI enables tailored care plans based on real-time patient data.	Overly data-driven decisions may overlook patient preferences or holistic needs.	Incorporate patient input alongside AI insights in care planning
Equitable access	Ensuring equitable access to high-quality care, regardless of socio-economic or geographic factors.	Remote monitoring expands access to care for patients in remote areas.	Patients without access to technology may face exclusions.	Offer alternative care options for patients without devices or internet access.
Transparent care	Being transparent about care decisions and maintaining open communication with patients and families.	AI decision logs enhance transparency in how care decisions are made.	Patients may struggle to understand complex AI processes.	Provide simplified explanations of AI processes during patient consultations.
Empathetic interactions	Demonstrating empathy and care in all interactions with patients and their families.	AI assists in identifying patient distress or changes in emotional well-being.	Automation may reduce the perceived empathy in care delivery.	Train staff to balance AI insights with empathetic human interactions.
Community outreach	Extending services to under-served communities through partnerships or subsidies.	AI reduces disparities by reaching patients in remote areas through wearable tech.	Privacy concerns may arise in under-served communities with less awareness of data usage.	Educate communities on data privacy and secure informed consent.

continued

Table A.1. continued

Core value	Description	Potential positive AI impacts	Potential negative AI impacts	Mitigations
Respect for choice	Respecting patient autonomy by allowing them to make informed decisions about their care.	AI empowers patients to make decisions based on real-time health data insights.	Patients may feel pressured to follow AI recommendations over personal preferences.	Educate patients on balancing AI insights with their personal care choices.
Avoiding harm	Ensuring no harm comes to patients through rigorous training and quality assurance.	AI detects health risks early, preventing harm through timely interventions.	Incorrect AI alerts could lead to unnecessary treatments or anxiety.	Validate alerts with clinical expertise before acting.
Integrity	Ensuring ethical practices in patient care and avoiding conflicts of interest in service delivery.	Transparent AI decisions reinforce ethical care practices.	Potential bias in AI algorithms may lead to unequal treatment.	Regularly audit and update AI systems to ensure fairness and address biases.

Table A.2. Risk analysis of potential negative impacts of AI.

Risk	**Likelihood**	**Severity**	**Initial risk level**	**Mitigation**	**Residual risk level**
Overly data-driven decisions may overlook patient preferences or holistic needs.	Medium	Medium	**Medium**	Incorporate patient input alongside AI insights.	**Low**
Patients without access to technology may face exclusions.	High	Medium	**High**	Offer alternative care options for patients without technology.	**Medium**
Patients may struggle to understand complex AI processes.	Medium	Low	**Low**	Provide simplified explanations during consultations.	**Low**
Automation may reduce the perceived empathy in care delivery.	Medium	Medium	**Medium**	Train staff to balance AI insights with empathetic interactions.	**Low**
Privacy concerns in underserved communities regarding data usage.	Medium	Medium	**Medium**	Educate communities on data privacy and secure informed consent.	**Low**
Patients may feel pressured to follow AI recommendations over preferences.	Medium	Medium	**Medium**	Educate patients on balancing AI insights with personal choices.	**Low**
Incorrect AI alerts could lead to unnecessary treatments or anxiety.	Low	Medium	**Low**	Validate alerts with clinical expertise before acting.	**Low**
Potential bias in AI algorithms may lead to unequal treatment.	Medium	High	**High**	Regularly audit and update AI systems to ensure fairness.	**Medium**

Based on this analysis, CareHouse decides that a partnership with NeuroAI is not misaligned with its values and so continues to explore a strategy for partnering.

CAREHOUSE'S AI STRATEGY

Strategic posture

CareHouse's strategic posture is best described as that of an *AI adaptor*. The company is not simply watching from the sidelines; it is actively embedding AI into its operations, partnering with NeuroAI to explore remote monitoring. At the same time, CareHouse is not aiming to disrupt their business model or the broader healthcare sector completely – at least not yet.

The approach the organisation takes is measured, values led, and grounded in augmenting the human touch rather than replacing it. That said, if the AI initiatives lead to scalable innovations in predictive care, CareHouse may well shift from being an AI adaptor to full-fledged *AI disruptor* in the future.

What problem is AI solving?

CareHouse's initial AI use case was remote patient monitoring, as this is a speciality of NeuroAI. However, to ensure it is picking the right solution, CareHouse stepped back to consider its strategic business goals and look for other places where AI could help. Table A.3 summarises its analysis of possible AI use cases, their importance and contribution to strategic business goals and their feasibility.

Strategy execution

Having chosen its AI Use Cases (Table A.3), CareHouse considers its approach to execution. CareHouse's execution strategy relies on embedding AI into its broader transformation agenda, rather than creating a stand-alone AI team. With a lean structure, a

Table A.3. CareHouse's strategic business goals and AI's contribution.

Strategic goal	**Problem it is trying to address**	**Potential AI use case**	**Will the use case solve the problem?**	**Is the use case feasible? Can we deliver it? What is the return on investment?**
Improve patient outcomes	Patients being readmitted due to missed signs of deterioration	Use AI-powered remote monitoring (via NeuroAI) to detect early signs of clinical risk	Yes, if data quality is sufficient and alerts are clinically validated. Needs integration with care pathways.	Medium feasibility (partnership in place); high ROI through avoided hospital admissions, better outcomes and improved reputation
Increase workforce efficiency	Nurses overwhelmed with administrative tasks like visit notes	Use generative AI to summarise and draft clinical notes based on structured visit data	Yes, particularly for repetitive documentation. May still require human review for complex cases.	High feasibility; medium-to-high ROI through reclaimed nurse time and improved morale
Enhance patient experience	Patients don't understand complex discharge instructions	Use AI to generate personalised, plain-English discharge summaries	Yes, especially for patients with low health literacy. Could improve compliance and satisfaction.	High feasibility; medium ROI through improved engagement and reduced call-backs or follow-ups
Reduce operational inefficiency	Inefficient nurse rostering and high rates of missed appointments	Use AI to optimise scheduling based on geography, risk levels and visit history	Likely, especially if paired with real-time updates and predictive no-show modelling.	Medium feasibility (requires data clean-up and rostering integration); high ROI from reduced travel time and improved service coverage
Improve access to care in rural areas	Delayed escalation for remote patients	Apply AI to wearable sensor data to trigger virtual check-ins	Yes, though only for patients willing/able to wear devices and share data.	Medium feasibility; high ROI in terms of avoided escalations and equity of care

single AI lead coordinates activity across clinical, operational, and technology domains. The company leverages external partnerships (like their collaboration with NeuroAI) for advanced capabilities, while focusing internal efforts on co-design, explainability and frontline adoption. Early success should come from quick-win use cases (like note summarisation), but also gradually building internal capability and AI literacy to support sustainable growth.

Table A.4 summarises CareHouse's approach for executing on AI strategy and its rationale.

Table A.4. CareHouse's strategic execution approach.

Approach	Fit for CareHouse?	Rationale
Embed AI in transformation	☑ Strong fit	AI should be woven into broader efforts to improve care delivery, not treated as a side project.
Use external partners	☑ Strong fit	CareHouse's partnership with NeuroAI is central; it should continue to partner for complex needs.
Co-design with end users	☑ Non-negotiable	Staff trust and patient safety are critical; this is essential for uptake and success.
Appoint an AI lead	☑ Recommended	Even in a small organisation, someone needs accountability for coordination and risk oversight.
Start with quick wins	☑ Recommended	Early, low-risk wins (e.g. admin note summaries) help build momentum and internal support.
Invest in internal capability	☑ Medium priority	Over time, CareHouse should build internal literacy and perhaps a small in-house data/AI team.
Central innovation team	☑ Not ideal	Likely to be too costly for an organisation of CareHouse's size. Better to coordinate through the AI lead.
Federated model	☒ Too complex	CareHouse isn't big enough for a decentralised model. Could cause duplication or confusion.

GOVERNANCE

Now that CareHouse has defined a clear AI strategy with strong alignment to business goals and a well-articulated approach to execution, it is almost time to implement. But not quite. CareHouse gave some consideration to AI risks when deciding whether or not to partner with NeuroAI. However, a fuller analysis is required so that a proper governance framework can be introduced.

CareHouse's AI governance framework is built around four key pillars (Table A.5): an AI Oversight Panel, a lightweight risk and impact assessment checklist, a commitment to explainable AI in clinical settings, and a continuous monitoring process for all deployed tools. This approach ensures that innovation is balanced with responsibility. It is practical enough for a mid-sized organisation, yet robust enough to meet the ethical and clinical demands of healthcare. For CareHouse, if an AI tool can't be explained, it doesn't go live.

CareHouse considers the list of mitigations and management practices for managing risk that are in Chapter 12. Based on this, CareHouse implements a set of mitigations. Table A.6 highlights these for two of CareHouse's highest priority risks.

Table A.5. Key elements of CareHouse's governance framework.

Governance element	What it ensures	Why it fits CareHouse
AI oversight panel	All AI aligns with clinical, ethical and operational expectations	Scales up oversight without slowing innovation
Risk & Impact Assessment (RIA)	Risks are considered before AI is deployed	Lightweight, practical, and easy to apply organisation wide
Explainability requirement	AI supports, not undermines, clinician judgement	Builds trust and reduces fear of automation
Continuous monitoring framework	Issues are caught early and models stay relevant	Ensures real-world performance and accountability

Table A.6. CareHouse critical AI risks and mitigations.

Risk	Mitigation or management practice	Adaptation
Potential bias in AI algorithms leading to unequal treatment.	*Kill switch* Create failsafe mechanisms to isolate and disable AI components in case of ethical issues.	Allow patient opt-out and have non-AI backup plans in place.
	Ethics guardian Assign a team member as a 'Critical Friend' to identify potential ethical violations.	Issue an Expression of Interest to the patient community for a paid advisor to represent patients who will sit in project meetings and raise ethical issues.
	Transparency pulse Publish regular reports detailing ethical AI practices and outcomes.	Publish a ***Patient Insights Report*** that shares anonymised data on AI-driven health trends, decision-making transparency and the measures taken to safeguard patient privacy and improve outcomes.
	Speak up! Establish anonymous channels for raising ethical concerns.	Establish a ***Care Integrity Hotline*** for staff and patients to anonymously report concerns about AI recommendations, potential biases, or any ethical issues in remote monitoring.
	Law watch Stay informed about and comply with relevant AI laws globally.	Task a member of the Legal team to report regularly to senior management on any changes in relevant laws.

Risk	Mitigation or management practice	Adaptation
Patients without access to technology may face exclusions.	*Bias busters* Foster diversity of thought within project teams to minimise bias.	Analyse patient profiles and assign each project team member the role of representing one patient segment.
	What-if tests Develop scenarios to test how AI behaves in unusual cases and extreme conditions.	Create personas representing patient profiles and stress-test how remote monitoring might affect them.
	Values check Reassess corporate values to ensure alignment with ethical AI goals.	Review corporate values to ensure inclusion is emphasised.
	Open forum Host public forums to gather input on AI-related ethical concerns.	Host ***Community Care Dialogues*** – virtual or in-person forums where patients, care-givers, and healthcare professionals can provide feedback and discuss ethical concerns about AI-driven care.

ESG

With AI evolving rapidly, CareHouse wants to understand how it affects the company's ESG performance. It starts with the previous year's ESG report (Table A.7, first two columns) and uses it to assess how each metric might be helped or harmed by AI (columns three and four). Where AI creates challenges, CareHouse defines actions to reduce negative impacts and then adjusts its ESG targets if needed (final column). In some cases, AI helps the company meet goals faster; in others, it needs to set more realistic targets.

Table A.7. CareHouse's assessment of how its AI adoption impacts ESG.

ESG category	Goal, initiatives and metrics	AI positive impacts	AI negative impacts and mitigations	Changes to metrics due to AI
Environmental				
GHG emissions	**Goal:** Achieve carbon neutrality by 2035. **Initiatives:** Transition contractor fleet to electric vehicles; promote telehealth services to reduce travel. **Metric:** Reduce travel-related emissions by 15% next financial year (FY).	Remote patient monitoring reduces the need for in-person visits, lowering emissions from transportation.	Increased energy use for real-time AI data processing. **Mitigation:** Partner with cloud providers using 100% renewable energy.	**Metric Improved:** Reduce travel-related emissions by 25% next FY due to AI-enabled remote patient monitoring.
Resource efficiency	**Goal:** Optimise material usage and minimise waste. **Initiatives:** Improve inventory management; increase use of biodegradable medical supplies. **Metric:** Use 50% biodegradable medical supplies next FY.	AI reduces unnecessary medical visits, leading to less use of disposable supplies.	Wearable devices generate electronic waste. **Mitigation:** Introduce an e-waste recycling program for returned devices.	**Metric Improved:** Use 60% biodegradable medical supplies next FY due to fewer disposable supplies needed during visits.

continued

Table A.7. continued

ESG category	Goal, initiatives and metrics	AI positive impacts	AI negative impacts and mitigations	Changes to metrics due to AI
Ecosystem impact	**Goal:** Maintain eco-friendly waste management and reduce environmental harm. **Initiatives:** Optimise waste collection logistics; conduct life-cycle assessments for medical devices. **Metric:** Divert 75% of non-hazardous waste from landfills next FY.	Reduced medical waste from fewer in-home visits.	Life-cycle impact of wearable devices, including manufacturing and disposal, creates environmental risks. **Mitigation:** Launch take-back programs for end-of-life devices.	**Metric Improved:** Divert 80% of non-hazardous waste from landfills next FY, aided by reduced waste during visits and improved recycling of electronic devices.
Social				
Diversity, equity and inclusion (DEI)	**Goal:** Rebalance CareHouse's aging workforce and increase age diversity **Initiatives:** Market opportunities to the 20 to 30-year-old sector; provide development opportunities for all ages. **Metric:** Improve age diversity, measured by coefficient of variation, to within 20% of optimal level within five years.	Use of AI appeals to younger generations who are more tech savvy and to those with young families who don't want to travel.	AI may disproportionately favour men and/or exclude older populations. **Mitigation:** Regularly audit hiring statistics to ensure equitable practices.	**Metric Improved:** Achieve coefficient of variation targets within three years.

ESG category	Goal, initiatives and metrics	AI positive impacts	AI negative impacts and mitigations	Changes to metrics due to AI
Human rights	**Goal:** Ensure fair treatment of contractor agencies providing temporary workers. **Initiatives:** Perform supplier audits; adopt an anti-modern slavery policy. **Metric:** Audit 90% of suppliers for human rights compliance next FY.	Remote patient monitoring reduces the need for temporary workers.	Remote monitoring reduces visibility of possible human rights issues among patients. **Mitigation:** Adopt a policy that ensures minimum annual face-time for patients.	**Metric Maintained:** Audit 90% of suppliers for human rights compliance next FY.
Labour management	**Goal:** Ensure fair pay and a balanced workload. **Initiatives:** Optimise schedules for contractors; introduce wellness programs. **Metric:** Maintain 85% contractor satisfaction next FY.	AI reduces workload and increasing satisfaction by minimising unnecessary visits, allowing contractors to focus on complex cases.	Adjustment challenges and stress due to learning new AI systems. **Mitigation:** Provide comprehensive training and ongoing technical support.	**Metric Weakened:** Expect a dip in contractor satisfaction to 75% as staff adjust to new AI systems.
Customer and community	**Goal:** Expand access to healthcare in underserved areas. **Initiatives:** Deliver free health screenings and workshops. **Metric:** Deliver 10,000 hours of free community healthcare services next FY.	Remote monitoring expands access to care for rural patients, increasing service delivery.	Lower face-to-face community interaction may weaken local relationships. **Mitigation:** Continue in-person workshops to maintain community presence.	**Metric Improved:** Deliver 12,000 hours of free community healthcare services next FY due to AI-enabled scalability.

continued

Table A.7. continued

ESG category	Goal, initiatives and metrics	AI positive impacts	AI negative impacts and mitigations	Changes to metrics due to AI
Cybersecurity and data privacy	**Goal:** Maintain full compliance with patient data protection laws. **Initiatives:** Use encrypted systems; conduct regular audits. **Metric:** Achieve 100% compliance with data protection laws next FY.	Use of AI for remote monitoring provides better documentation, making compliance easier to track.	Increased data volume from remote monitoring heightens breach risks. **Mitigation:** Partner with cybersecurity firms to enhance protection and monitor risks.	**Metric Weakened:** Achieve 98% compliance next FY due to expected teething issues with AI systems.
Health and safety	**Goal:** Reduce workplace injuries and improve contractor safety. **Initiatives:** Enhance safety protocols; provide safety training. **Metric:** Reduce workplace incidents by 15% next FY.	Remote monitoring reduces on-site risks by lowering the need for frequent home visits.	Safety incidents from wearable device malfunctions. **Mitigation:** Regular device testing and training for proper usage.	**Metric Maintained:** Reduce workplace incidents by 15% next FY.
Governance				
Board and management	**Goal:** Ensure diverse and accountable leadership. **Initiatives:** Implement term limits and diversity targets. **Metric:** Maintain 30% representation of women and minorities on the board.	AI makes patient visits more efficient, thus reducing fire-fighting on the board, leading to more strategic discussions.	Lack of AI knowledge on the board leads to poor oversight. **Mitigation:** Mandatory AI training for all board members.	**Metric Maintained:** Maintain 30% representation of women and minorities on the board next FY.

ESG category	Goal, initiatives and metrics	AI positive impacts	AI negative impacts and mitigations	Changes to metrics due to AI
Policy	**Goal:** Develop sustainability policies. **Initiatives:** Update ESG policies annually. **Metric:** Implement and regularly ESG policy.	Introduce AI ethics board, which enhances compliance with ethical standards in patient care and decision-making.	Rapid AI advancements create policy gaps if not regularly updated. **Mitigation:** Conduct annual reviews of AI ethics policies.	**Metric Updated:** Implement and regularly update AI ethics and ESG policies to address emerging issues.
Disclosure and reporting	**Goal:** Publish comprehensive ESG reports annually. **Initiatives:** Host stakeholder meetings; enhance transparency in reporting. **Metric:** Publish comprehensive ESG reports annually.	Disclosure of AI use helps to build trust in the company.	Complex AI data models may reduce accessibility for lay audiences. **Mitigation:** Simplify reporting language and provide visual summaries of AI data.	**Metric Updated:** Publish comprehensive ESG reports annually with added AI-focused metrics.

References

1. Chui M, Hazan E, Roberts R, Singla A, Smaje K, Sukharevsky A, Yee L, Zemmel R (2023) *The Economic Potential of Generative AI: The Next Productivity Frontier.* McKinsey & Company, New York, NY.
2. Fioretti L, La Croce C, Siviero A, Clemmons E (2024) *The Global Impact of Artificial Intelligence on the Economy and Jobs: AI Will Steer 3.5% of GDP in 2030.* International Data Corporation, Needham, MA.
3. Hajkowicz S, Whittle J (2023) Yes, AI could help us fix the productivity slump – but it can't fix everything. *The Conversation*, 22 June. Available at: <https://theconversation.com/yes-ai-could-help-us-fix-the-productivity-slump-but-it-cant-fix-everything-207623> (accessed 1 June 2025).
4. Ryseff J, De Bruhl BF, Newberry SJ (2024) *The Root Causes of Failure for Artificial Intelligence Projects and How They Can Succeed: Avoiding the Anti-Patterns of AI.* RAND Corporation, Santa Monica, CA.
5. de Bellefonds N, Charanya T, Franke MR, Apotheker J, Forth P, Grebe M, Luther A, de Laubier R, Lukic V, Martin M, Nopp C, Sassine J (2024) *Where's the Value in AI?* Boston Consulting Group, Boston, MA.
6. Maslej N, Fattorini L, Perrault R, Gil Y, Parli V, Kariuki N, Capstick E, Reuel A, Brynjolfsson E, Etchemendy J, Ligett K, Lyons T, Manyika J, Niebles JC, Shoham Y, Wald R, Walsh T, Hamrah A, Santarlasci L, Betts Lotufo J, Rome A, Shi A, Oak S (2025) *Artificial Intelligence Index Report 2025.* Stanford Institute for Human-Centered Artificial Intelligence, Stanford, CA. Available at: <https://hai.stanford.edu/ai-index/2025-ai-index-report> (accessed 1 June 2025).
7. Semrush (2025) *Most visited websites in the world.* Semrush, Boston, MA. Available at: <https://www.semrush.com/website/top/?utm_source=explodingtopics.com&utm_medium=referral&utm_campaign=chatgpt-users> (accessed 1 June 2025).
8. Turing AM (1950) Computing machinery and intelligence. *Mind* **59**, 433–460.

9. Mei Q, Xie Y, Yuan W, Jackson MO (2024) A Turing test of whether AI chatbots are behaviorally similar to humans. *Proceedings of the National Academy of Sciences of the United States of America* **121**(9), e2313925121. doi:10.1073/pnas.2313925121
10. Rosenblatt F (1958) The perceptron: a probabilistic model for information storage and organization in the brain. *Psychological Review* **65**(6), 386–408.
11. ALPAC (1966) *Languages and Machines: Computers in Translation and Linguistics*. Automatic Language Processing Advisory Committee, Division of Behavioral Sciences, National Academy of Sciences, National Research Council, Washington, DC.
12. Minsky M, Papert S (1969) *Perceptrons: An Introduction to Computational Geometry*. MIT Press, Cambridge, MA.
13. Lighthill J (1973) Artificial intelligence: A general survey. In *Artificial Intelligence: A Paper Symposium*. Science Research Council, London. <doi:10.1016/0004-3702(74)90016-2>
14. Kennedy B, Tyson A, Saks E (2023) Public awareness of artificial intelligence in everyday activities. Pew Research Center. Available at: <https://www.pewresearch.org/science/2023/02/15/public-awareness-of-artificial-intelligence-in-everyday-activities/> (accessed 1 June 2025).
15. Tech Business News (2025) 402.74 million terabytes of data is created every day. *Tech Business News*. Available at: <https://www.techbusinessnews.com.au/blog/402-74-million-terrabytes-of-data-is-created-every-day/> (accessed 1 June 2025).
16. Marr B (2018) How much data do we create every day? The mind-blowing stats everyone should read. *Forbes*. Available at: <https://www.forbes.com/sites/bernardmarr/2018/05/21/how-much-data-do-we-create-every-day-the-mind-blowing-stats-everyone-should-read/> (accessed 1 June 2025).
17. Jaffri A (2024) Hype cycle for artificial intelligence 2024. Gartner. Available at: <https://www.gartner.com/en/articles/hype-cycle-for-artificial-intelligence> (accessed 1 June 2025).
18. Adapted from Kozyrkov C (2019) Why AI is here to stay. HackerNoon. Available at: <https://hackernoon.com/why-ai-is-here-to-stay-9c75b1868b9b> (accessed 1 June 2025).
19. Wolfram S (2023) What is ChatGPT doing ... and why does it work? Stephen Wolfram Writings. Available at: <https://writings.

stephenwolfram.com/2023/02/what-is-chatgpt-doing-and-why-does-it-work/> (accessed 1 June 2025).

20. Singla A, Sukharevsky A, Yee L, Chui M, Hall B (2025) The state of AI: How organizations are rewiring to capture value. McKinsey & Company, New York, NY. Available at: <https://www.mckinsey.com/capabilities/quantumblack/our-insights/the-state-of-ai> (accessed 1 June 2025).
21. Gartner (2023) The CIO's guide to artificial intelligence. Gartner. Available at: <https://www.gartner.com/smarterwithgartner/the-cios-guide-to-artificial-intelligence> (accessed 1 June 2025).
22. MITRE (2023) The MITRE AI Maturity Model and Organizational Assessment Tool Guide: A path to successful AI adoption. MITRE Corporation. Available at: <https://www.mitre.org/news-insights/publication/mitre-ai-maturity-model-and-organizational-assessment-tool-guide> (accessed 1 June 2025).
23. Travel BBC (2024) Air Canada chatbot misinformation: What travellers should know. BBC Travel. Available at: <https://www.bbc.com/travel/article/20240222-air-canada-chatbot-misinformation-what-travellers-should-know> (accessed 1 June 2025).
24. Thomson J (2025) Top CEOs reveal 17 ways to make Australia great again. *Australian Financial Review*, 24 January. Available at: <https://www.afr.com/chanticleer/top-ceos-reveal-17-ways-to-make-australia-great-again-20250124-p5l6w3> (accessed 1 June 2025).
25. Berghel H (2018) Malice domestic: the Cambridge Analytica dystopia. *Computer* **51**(5), 84–89.
26. Lavoipierre A (2024) Bunnings breached privacy laws by using facial recognition on customers, Commissioner finds. ABC News. Available at: <https://www.abc.net.au/news/2024-11-19/oaic-investigation-into-bunnings-facial-recognition/104613700> (accessed 1 June 2025).
27. Briggs J, Kodnani D (2023) The potentially large effects of artificial intelligence on economic growth. Goldman Sachs, New York, NY. Available at: <https://www.gspublishing.com/content/research/en/reports/2023/03/27/d64e052b-0f6e-45d7-967b-d7be35fabd16.html> (accessed 1 June 2025).
28. Chui M, Roberts R, Yee L, Hazan E, Singla A, Sukharevsky A, Smaje K, Zemmel R (2023) The economic potential of generative AI: The next productivity frontier. McKinsey & Company, New York, NY.

Available at: <https://www.mckinsey.com/capabilities/mckinsey-digital/our-insights/the-economic-potential-of-generative-ai-the-next-productivity-frontier> (accessed 1 June 2025).

29. Jaokar A (2024) The AI radiologists replacement saga: Don't be misled by the scaremongering – science vs. science fiction. Data Science Central, London. Available at: <https://www.datasciencecentral.com/the-ai-radiologists-replacement-saga-dont-be-misled-by-the-scaremongering-science-v-s-science-fiction/> (accessed 1 June 2025).
30. Frey CB, Osborne MA (2013) The future of employment: how susceptible are jobs to computerisation? Working Paper, Oxford Martin School, University of Oxford, Oxford. Available at: <https://www.oxfordmartin.ox.ac.uk/publications/the-future-of-employment> (accessed 1 June 2025).
31. Pessimists Archive (n.d.) <https://pessimistsarchive.org/> (accessed 1 June 2025).
32. Schrader D (2023) A retrospective on AI predictions over the decades. LinkedIn. Available at: <https://www.linkedin.com/pulse/retrospective-ai-predictions-over-decades-shiftavenue/> (accessed 1 June 2025).
33. Nazneen M, Sgroi D (2023) Happiness and productivity in the workplace. In *Oxford Research Encyclopedia of Economics and Finance* (Ed. Anindya Banerjee). Oxford University Press, Oxford.
34. Brynjolfsson E, Li D, Raymond LR (2023) Generative AI at work. NBER Working Paper No. 31161. National Bureau of Economic Research, Cambridge, MA.
35. Dell'Acqua F, McFowland E, III, Mollick E, Lifshitz-Assaf H, Kellogg KC, Rajendran S, Krayer L, Candelon F, Lakhani KR (2023) Navigating the jagged technological frontier: field experimental evidence of the effects of AI on knowledge worker productivity and quality. Harvard Business School Working Paper No. 24-013, Harvard, MA.
36. Dell'Acqua F, Ayoubi C, Lifshitz-Assaf H, Sadun R, Mollick ER, Mollick L, Han Y, Goldman J, Nair H, Taub S, Lakhani KR (2025) The cybernetic teammate: a field experiment on generative AI reshaping teamwork and expertise. Harvard Business School Working Paper No. 25-043, Harvard, MA.
37. Peng S, Kalliamvakou E, Cihon P, Demirer M (2023) The impact of AI on developer productivity: evidence from GitHub Copilot. *arXiv*.

Available at: <https://arxiv.org/abs/2302.06590> (accessed 9 June 2025).

38. Cui Z, Demirer M, Jaffe S, Musolff L, Peng S, Salz T (2025) The effects of generative AI on high-skilled work: evidence from three field experiments with software developers. *SSRN*. Available at: <https://papers.ssrn.com/sol3/papers.cfm?abstract_id=4945566> (accessed 9 June 2025).
39. Gambacorta L, Qiu H, Shan S, Rees D (2024) Generative AI and labour productivity: a field experiment on coding. BIS Working Paper No. 1208. Bank for International Settlements, Basel. Available at: <https://www.bis.org/publ/work1208.htm> (accessed 9 June 2025).
40. Kalliamvakou E, Peng S, Cihon P, Demirer M (2023) Research: quantifying GitHub Copilot's impact on developer productivity and happiness. GitHub Blog. Available at: <https://github.blog/news-insights/research/research-quantifying-github-copilots-impact-on-developer-productivity-and-happiness/> (accessed 9 June 2025).
41. Monahan K, Burlacu G (2024) From burnout to balance: AI-enhanced work models for the future. Upwork Research Institute, Palo Alto, CA. Available at: <https://www.upwork.com/research/ai-enhanced-work-models> (accessed 9 June 2025).
42. Lockett W (2024) Intel admits AI decreases productivity. *Predict (Medium)*. Available at: <https://medium.com/predict/intel-admits-ai-decreases-productivity-226681d1af18> (accessed 9 June 2025).
43. Bano M, Zowghi D, Whittle J, Zhu L, Reeson A, Martin R, Parsons J (2024) Survey insights on M365 Copilot adoption. *arXiv*. Available at: <https://arxiv.org/abs/2412.16162> (accessed 9 June 2025).
44. Digital Transformation Agency (2024) Evaluation of whole-of-government trial into generative AI: now available. Digital Transformation Agency, Canberra. Available at: <https://www.dta.gov.au/blogs/evaluation-whole-government-trial-generative-ai-now-available> (accessed 9 June 2025).
45. Kabir S, Udo-Imeh DN, Kou B, Zhang T (2024) Is stack overflow obsolete? An empirical study of the characteristics of ChatGPT answers to stack overflow questions. In *Proceedings of the 2024 CHI Conference on Human Factors in Computing Systems* Article 935, pp. 1–17.
46. Filippucci F, Gal P, Jona-Lasinio C, Leandro A, Nicoletti G (2024) *The Impact of Artificial Intelligence on Productivity, Distribution and*

Growth: Key Mechanisms, Initial Evidence and Policy Challenges. OECD Artificial Intelligence Papers, No. 15. OECD Publishing, Paris.

47. Siedschlag I, Duran Vanegas JD (2024) Does artificial intelligence enhance firm productivity? *SSRN*. Available at: <https://ssrn.com/abstract=4974850> (accessed 9 June 2025).
48. Lockett W (2024) Intel admits AI decreases productivity. *Predict (Medium)*. Available at: <https://medium.com/predict/intel-admits-ai-decreases-productivity-226681d1af18> (accessed 9 June 2025).
49. Roberts M, Driggs D, Thorpe M, *et al.* (2021) Common pitfalls and recommendations for using machine learning to detect and prognosticate for COVID-19 using chest radiographs and CT scans. *Nature Machine Intelligence* **3**, 199–217.
50. Grudin J (1988) Why CSCW applications fail: problems in the design and evaluation of organizational interfaces. In *Proceedings of the 1988 ACM Conference on Computer-Supported Cooperative Work*, Portland, OR, 26–28 September 1988. pp. 85–93. ACM, New York, NY.
51. Business Council of Australia (2024) *Australia's flagging productivity and international competitiveness.* Business Council of Australia, 19 November 2024. Available at: <https://assets.nationbuilder.com/bca/pages/9204/attachments/original/1731995892/Australia%E2%80%99s_flagging_productivity_and_international_competitiveness_20241119_WEB.pdf> (accessed 9 June 2025).
52. Eurostat (2025) Productivity trends using key national accounts indicators. *Eurostat Statistics Explained.* Available at: https://ec.europa.eu/eurostat/statistics-explained/index.php?title=Productivity_trends_using_key_national_accounts_indicators (accessed 9 June 2025).
53. Bureau of Communications and Arts Research (2017) Digital productivity: key issues from the literature. Department of Communications and the Arts, Canberra. Available at: <https://www.infrastructure.gov.au/sites/default/files/documents/digital-productivity-key-issues-from-the-literature_0.pdf> (accessed 9 June 2025).
54. Macdonald S, Anderson P, Kimbel D (2000) Measurement or management? Revisiting the productivity paradox of information technology. *Vierteljahrshefte zur Wirtschaftsforschung* **69**, 601–617.
55. Brynjolfsson E, Rock D, Syverson C (2018) AI and the modern productivity paradox: a clash of expectations and statistics. *MIT Initiative on the Digital Economy Research Brief* 2018.01. Available

at: <https://ide.mit.edu/sites/default/files/publications/IDE%20 Research%20Brief_v0118.pdf> (accessed 9 June 2025).

56. McConnon A (2025) DeepSeek's reasoning AI shows power of small models, efficiently trained. *IBM Think*, 27 January 2025. Available at: <https://www.ibm.com/think/news/deepseek-r1-ai> (accessed 9 June 2025).
57. Olavsrud T (2025) 12 famous AI disasters. *CIO*, 9 June 2025. Available at: <https://www.cio.com/article/190888/5-famous-analytics-and-ai-disasters.html> (accessed 10 June 2025).
58. Purtill J (2024) Cosmos Magazine publishes AI-generated articles, drawing criticism from journalists, co-founders. *ABC News*, 8 August 2024. Available at: <https://www.abc.net.au/news/science/2024-08-08/csiro-cosmos-magazine-generating-articles-using-ai/104186330> (accessed 10 June 2025).
59. Rome S, Chen T, Tang R, Zhou L, Ture F (2024) 'Ask me anything': how Comcast uses LLMs to assist agents in real time. In *Proceedings of the 47th International ACM SIGIR Conference on Research and Development in Information Retrieval (SIGIR '24)*, Washington, DC, 14–18 July 2024, pp. 2827–2831. Association for Computing Machinery, New York, NY.
60. Kovoor JG, Bacchi S, Gupta AK, Stretton B, Malycha J, Reddi BA, Liew D, O'Callaghan PG, Beltrame JF, Zannettino AC, Jones KL, Horowitz M, Dobbins C, Hewett PJ, Trochsler MI, Maddern GJ (2023) The Adelaide Score: an artificial intelligence measure of readiness for discharge after general surgery. *ANZ Journal of Surgery* **93**, 2119–2124.
61. You M, Wang Q, Sun H, Castro I, Jiang J (2022) Digital twins based day-ahead integrated energy system scheduling under load and renewable energy uncertainties. *Applied Energy* **305**, 117899.
62. Ziyad M, Tjandra K, Zulvah, Sugihartanto MF, Arief M (2022) An optimized and safety-aware maintenance framework: a case study on aircraft engine. In *Proceedings of the 2022 IEEE 25th International Conference on Intelligent Transportation Systems (ITSC)*, Macau, China, 8–12 October 2022, pp. 2057–2062. IEEE, London.
63. Durongkadej I, Hu W, Wang HE (2024) How artificial intelligence incidents affect banks and financial services firms? A study of five firms. *Finance Research Letters* **70**, 106279.
64. Johnson K (2024) California's two biggest school districts botched AI deals. Here are lessons from their mistakes. *The Markup*, 6 August

2024. Available at: <https://themarkup.org/artificial-intelligence/2024/08/06/californias-two-biggest-school-districts-botched-ai-deals-here-are-lessons-from-their-mistakes> (accessed 10 June 2025).

65. Houser K (2019) Investor sues after an AI's automated trades cost him $20 million. *Futurism*, 6 May 2019. Available at: <https://futurism.com/investing-lawsuit-ai-trades-cost-millions> (accessed 10 June 2025).
66. Fernandez-Macias E, Klenert D, Antón JI (2020) Don't blame it on the machines: robots and employment in Europe. *VoxEU.org*, 24 February 2020. Available at: <https://cepr.org/voxeu/columns/dont-blame-it-machines-robots-and-employment-europe> (accessed 10 June 2025).
67. Acemoglu D, Restrepo P (2020) Robots and jobs: evidence from US labor markets. *Journal of Political Economy* **128**(6), 2188–2244.
68. U.S. Bureau of Labor Statistics (2025) Computer and information research scientists. *Occupational Outlook Handbook*. Available at: <https://www.bls.gov/ooh/computer-and-information-technology/computer-and-information-research-scientists.htm> (accessed 10 June 2025).
69. Simplilearn (2025) AI engineer salary in the US: insights and average earnings. *Simplilearn*. Available at: <https://www.simplilearn.com/ai-engineer-salary-in-us-article> (accessed 10 June 2025).
70. World Economic Forum (2020) The future of jobs report 2020. World Economic Forum, Geneva. Available at: <https://www.weforum.org/publications/the-future-of-jobs-report-2020/> (accessed 10 June 2025).
71. Abel JR, Deitz R, Emanuel N, Hyman B (2024) AI and the labor market: will firms hire, fire, or retrain? *Liberty Street Economics*, 4 September 2024. Available at: <https://libertystreeteconomics.newyorkfed.org/2024/09/ai-and-the-labor-market-will-firms-hire-fire-or-retrain/> (accessed 10 June 2025).
72. Broadband TV News (2025) Sky to cut 2,000 call centre jobs amid AI shift. *Broadband TV News*, 28 March 2025. Available at: <https://www.broadbandtvnews.com/2025/03/28/sky-to-cut-2000-call-centre-jobs-amid-ai-shift/> (accessed 10 June 2025).
73. Butler S (2025) Ocado to cut 500 technology and finance jobs as AI reduces costs. *The Guardian*, 27 February 2025. Available at: <https://www.theguardian.com/business/2025/feb/27/ocado-to-cut-500-technology-and-finance-jobs-as-ai-reduces-costs> (accessed 10 June 2025).

74. Kunert P (2025) Workday erases 8.5% of workforce because of AI. *The Register*, 5 February 2025. Available at: <https://www.theregister.com/2025/02/05/workday_restructure_job_cuts/> (accessed 10 June 2025).
75. O'Sullivan I (2025) Companies that have replaced workers with AI in 2024 & 2025. *Tech.co*, 16 May 2025. Available at: <https://tech.co/news/companies-replace-workers-with-ai> (accessed 10 June 2025).
76. Huang H (2023) What CEOs need to know about the costs of adopting GenAI. *Harvard Business Review* 15. Available at: <https://hbr.org/2023/11/what-ceos-need-to-know-about-the-costs-of-adopting-genai> (accessed 11 August 2025).
77. Buchholz K (2024) The extreme cost of training AI models. *Forbes*, 23 August 2024. Available at: <https://www.forbes.com/sites/katharinabuchholz/2024/08/23/the-extreme-cost-of-training-ai-models/> (accessed 10 June 2025).
78. Buzby H (2024) Breaking down the cost of large language models. *Qwak*, 25 June 2024. Available at: <https://www.qwak.com/post/llm-cost> (accessed 10 June 2025).
79. IBM Institute for Business Value (2024) The CEO's guide to generative AI: cost of compute. IBM, Armonk, NY. Available at: <https://www.ibm.com/thought-leadership/institute-business-value/report/ceo-generative-ai/ceo-ai-cost-of-compute> (accessed 10 June 2025).
80. Howarth J (2025) ChatGPT Enterprise pricing, features and limitations. *Exploding Topics*, 28 May 2025. Available at: <https://explodingtopics.com/blog/chatgpt-enterprise> (accessed 10 June 2025).
81. Savitsky Z (2024) 'Brutal' math test stumps AI but not human experts. *Science*, 3 December 2024. Available at: <https://www.science.org/content/article/brutal-math-test-stumps-ai-not-human-experts> (accessed 10 June 2025).
82. Eaton K (2024) How many R's in 'strawberry'? This AI doesn't know. *Inc.*, 28 August 2024. Available at: <https://www.inc.com/kit-eaton/how-many-rs-in-strawberry-this-ai-cant-tell-you.html> (accessed 10 June 2025).
83. Almquist E, Senior J, Bloch N (2016) The elements of value. *Harvard Business Review*. Available at: <https://hbr.org/2016/09/the-elements-of-value> (accessed 11 August 2025).
84. Schwartz SH (2012) An overview of the Schwartz theory of basic values. *Online Readings in Psychology and Culture* **2**(1), 11.

85. Frank EA (2019) Faster, better, cheaper: a maligned era of NASA's history. *Colliding Worlds*, 28 January 2019. Available at: <https://www.elizabethafrank.com/colliding-worlds/fbc> (accessed 10 June 2025).
86. DataReportal (2025) Global social media statistics. *DataReportal.* Available at: <https://datareportal.com/social-media-users> (accessed 10 June 2025).
87. Jumper J, Evans R, Pritzel A, Green T, Figurnov M, Ronneberger O, *et al.* (2021) Highly accurate protein structure prediction with AlphaFold. *Nature* **596**, 583–589.
88. Meier JD (2024) 10 best innovation frameworks. *JD Meier*, 28 May 2024. Available at: <https://jdmeier.com/10-best-innovation-frameworks/> (accessed 10 June 2025).
89. Herrmann Asia (2025) The HBDI®. *Herrmann Asia.* Available at: <https://herrmann.com.au/hbdi/> (accessed 10 June 2025).
90. Courtney H, Kirkland J, Viguerie P (1997) Strategy under uncertainty. *Harvard Business Review* November–December 1997. Available at: <https://hbr.org/1997/11/strategy-under-uncertainty> (accessed 11 August 2025).
91. Altrata Limited (2024) Private sector drives 63% of Chief AI Officer appointments as AI influences corporate talent strategy. *PR Newswire*, 12 December 2024. Available at: <https://www.prnewswire.com/news-releases/private-sector-drives-63-of-chief-ai-officer-appointments-as-ai-influences-corporate-talent-strategy-302330112.html> (accessed 10 June 2025).
92. Digital Transformation Agency (2024) Evaluation of whole-of-government trial into generative AI: now available. *Digital Transformation Agency*, 23 October 2024. Available at: <https://www.dta.gov.au/blogs/evaluation-whole-government-trial-generative-ai-now-available> (accessed 10 June 2025).
93. Gillespie N, Lockey S, Curtis C, Pool J, Akbari A (2023) *Trust in Artificial Intelligence: A Global Study.* The University of Queensland, Brisbane, and KPMG Australia.
94. Sikora T (2025) NACD quarterly survey: Q4 2024—preparing for a changing landscape. *National Association of Corporate Directors*, 24 January 2025. Available at: <https://www.nacdonline.org/all-governance/governance-resources/governance-surveys/surveys-benchmarking/nacd-quarterly-survey-q4-2024/> (accessed 10 June 2025).

95. Deloitte Insights (2024) Oversight of AI in the boardroom. *Deloitte Insights*, 8 November 2024. Available at: <https://www2.deloitte.com/us/en/insights/topics/leadership/successful-ai-oversight-may-require-more-engagement-in-the-boardroom.html> (accessed 10 June 2025).
96. Sherman N (2024) AI regulations around the world – 2025. *Mind Foundry*, 25 January 2024. Available at: <https://www.mindfoundry.ai/blog/ai-regulations-around-the-world> (accessed 10 June 2025).
97. Byrne D (2024) *Ten Steps to Creating an AI Policy*. The Corporate Governance Institute, Melbourne.
98. Australian Institute of Company Directors and Human Technology Institute (2024) *A Director's Guide to AI Governance*. Australian Institute of Company Directors, Sydney.
99. Organisation for Economic Co-operation and Development (OECD) (2025) OECD AI Policy Observatory. *OECD.AI*. Available at: <https://oecd.ai/en> (accessed 15 June 2025).
100. Etchegaray D, Luo Y, Li Y, Do B, Liu J, Huang Z, Kusy B (2024) Edge deployable online domain adaptation for underwater object detection. In *Proceedings of the 2024 International Joint Conference on Neural Networks (IJCNN)*, 30 June–5 July 2024, Yokohama, Japan, pp. 1–9. IEEE, London. <doi:10.1109/ijcnn60899.2024.10650705> (accessed 29 July 2025).
101. Garg S, Forbes-Smith N, Hilton JE, Prakash M (2018) SparkCloud: a cloud-based elastic bushfire simulation service. *Remote Sensing* **10**(1), 74.
102. St Vincent's Hospital Melbourne (2021) AI could be the key to transforming breast cancer screening. *St Vincent's Hospital Melbourne*, 2 August 2021. Available at: <https://www.svhm.org.au/newsroom/news/ai-could-be-the-key-to-transforming-breast-cancer-screening> (accessed 15 June 2025).
103. Ireland D, Atay C, Liddle J, Bradford D, Lee H, Rushin O, Mullins T, Angus D, Wiles J, McBride S, Vogel AP (2016) Hello Harlie: enabling speech monitoring through chat-bot conversations. In *Digital health innovation for consumers, clinicians, connectivity and community: Selected papers from the 24th Australian National Health Informatics Conference (HIC 2016)*, July 2016, Melbourne, Australia. (Eds A Georgiou, LK Schaper, and S Whetton). Available at: <https://pubmed.ncbi.nlm.nih.gov/27440289/> (accessed 29 July 2025).

104. Haidt J (2024) *The Anxious Generation: How the Great Rewiring of Childhood Is Causing an Epidemic of Mental Illness*. Penguin Press, New York, NY.
105. European Union (2024) Artificial Intelligence Act (Regulation (EU) 2024/1689). *Official Journal of the European Union*, L 2024/1689, 12 July 2024. Available at: <https://eur-lex.europa.eu/eli/reg/2024/1689/oj> (accessed 15 June 2025).
106. AI Ethicist (2025) Frameworks, guidelines, toolkits. *AI Ethicist*. Available at: <https://www.aiethicist.org/frameworks-guidelines-toolkits> (accessed 15 June 2025).
107. Lu Q, Zhu L, Whittle J, Xu X (2023) *Responsible AI: Best Practices for Creating Trustworthy AI Systems*. Addison-Wesley Professional, Boston, MA.
108. Various sources. For a thorough analysis, see Luccioni S, Jernite Y, Strubell E (2024) Power hungry processing: watts driving the cost of AI deployment? In *Proceedings of the 2024 ACM Conference on Fairness, Accountability, and Transparency (FAccT '24)*, New York, NY, pp. 85–99. Association for Computing Machinery, New York, NY. Available at: <https://arxiv.org/abs/2311.16863> (accessed 29 July 2025).
109. Tomlinson B, Black RW, Patterson DJ, Torrance AW (2024) The carbon emissions of writing and illustrating are lower for AI than for humans. *Scientific Reports* **14**, 3732.
110. Crawford K (2021) *Atlas of AI: Power, Politics, and the Planetary Costs of Artificial Intelligence*. Yale University Press, New Haven, CT.
111. Merchant B (2017) *The One Device: The Secret History of the iPhone*. Little, Brown and Company, New York, NY.
112. Apple Inc (2024) *Apple Environmental Progress Report 2024*. Apple Inc., Cupertino, CA. Available at: <https://www.apple.com/au/environment/pdf/Apple_Environmental_Progress_Report_2024.pdf> (accessed 15 June 2025).
113. Gómez M, Grimes S, Qian Y, Feng Y, Fowler G (2023) Critical and strategic metals in mobile phones: a detailed characterisation of multigenerational waste mobile phones and the economic drivers for recovery of metal value. *Journal of Cleaner Production* **419**, 138099.
114. Flaningam E (2024) A primer on data centers. *Generative Value*, 4 February. Available at: <https://www.generativevalue.com/p/a-primer-on-data-centers> (accessed 15 June 2025).

115. Flaningam E (2024) A primer on AI data centers. *Generative Value*, 13 October. Available at: <https://www.generativevalue.com/p/a-primer-on-ai-data-centers> (accessed 15 June 2025).
116. Gooding M (2024) Global data center electricity use to double by 2026 – IEA report. *Data Center Dynamics*, 26 January. Available at: <https://www.datacenterdynamics.com/en/news/global-data-center-electricity-use-to-double-by-2026-report/> (accessed 15 June 2025).
117. Deloitte (2024) *TMT Predictions 2025: Bridging the Gaps*. Deloitte Insights, 19 November. Available at: <https://www2.deloitte.com/content/dam/insights/articles/ca187597_tmtp25/DI_TMT-predictions-2025.pdf> (accessed 15 June 2025).
118. Ramachandran K, Stewart D, Hardin K, Crossan G, Bucaille A (2024) As generative AI asks for more power, data centers seek more reliable, cleaner energy solutions. *Deloitte Insights*, 19 November. Available at: <https://www2.deloitte.com/us/en/insights/industry/technology/technology-media-and-telecom-predictions/2025/genai-power-consumption-creates-need-for-more-sustainable-data-centers.html> (accessed 15 June 2025).
119. Li P, Yang J, Islam MA, Ren S (2024) Making AI less 'thirsty': uncovering and addressing the secret water footprint of AI models. *Communications of the ACM* (in press). Preprint available at: <https://arxiv.org/abs/2304.03271> (accessed 15 June 2025).
120. Young J (2024) Why AI is so thirsty: data centers use massive amounts of water. *Newsweek*, 24 March. Available at: <https://www.newsweek.com/why-ai-so-thirsty-data-centers-use-massive-amounts-water-1882374> (accessed 15 June 2025).
121. Marr B (2023) A short history of ChatGPT: how we got to where we are today. *Forbes*, 19 May. Available at: < https://www.forbes.com/sites/bernardmarr/2023/05/19/a-short-history-of-chatgpt-how-we-got-to-where-we-are-today/> (accessed 15 June 2025).
122. Young B (2023) AI expert speculates on GPT-4 architecture. *Weights & Biases*, 21 June. Available at: <https://wandb.ai/byyoung3/ml-news/reports/AI-Expert-Speculates-on-GPT-4-Architecture—VmlldzoONzAONzg4> (accessed 15 June 2025).
123. de Vries A (2023) The growing energy footprint of artificial intelligence. *Joule* 7, 2191–2194.

124. Desislavov R, Martínez-Plumed F, Hernández-Orallo J (2023) Trends in AI inference energy consumption: beyond the performance-vs-parameter laws of deep learning. *Sustainable Computing: Informatics and Systems* **38**, 100857.
125. Flensted T (2024) How many people use Google? Statistics & facts (2025). SEO.ai. Available at: <https://seo.ai/blog/how-many-people-use-google> (accessed 15 June 2025).
126. DatacenterDynamics (2023) IBM claims its data centers run on 74 percent renewable energy. Available at: <https://www.datacenterdynamics.com/en/news/ibm-claims-its-data-centers-run-on-74-percent-renewable-energy/> (accessed 15 June 2025).
127. Flanagan C (2023) How data centers are driving the renewable energy transition. *Forbes*. Available at: <https://www.forbes.com/sites/siemens-smart-infrastructure/2023/03/13/how-data-centers-are-driving-the-renewable-energy-transition/> (accessed 15 June 2025).
128. CSIRO (2024) Groundwater Cooling Project overview. CSIRO. Available at: <https://groundwatercooling.csiro.au/overview.html> (accessed 15 June 2025).
129. Law M (2024) Microsoft unveils zero-water cooling for AI data centres. *Data Centre Magazine.* Available at: <https://datacentremagazine.com/articles/microsoft-unveils-zero-water-cooling-for-ai-data-centres> (accessed 15 June 2025).
130. Kitsios V (2024) AI weather models can now beat the best traditional forecasts. *The Conversation.* Available at: <https://theconversation.com/ai-weather-models-can-now-beat-the-best-traditional-forecasts-245168> (accessed 15 June 2025).
131. Materia S, Palma García L, van Straaten C, Coumou D, de Luca P, Kretschmer M, Donat MG (2023) Artificial intelligence for climate prediction of extremes: state of the art, challenges, and future perspectives. *Wiley Interdisciplinary Reviews: Climate Change* **14**, e914.
132. CSIRO and Bureau of Meteorology (2024) *State of the Climate 2024.* CSIRO Publishing, Melbourne.
133. Australasian Fire and Emergency Service Authorities Council (AFAC) (2025) Fire prediction simulators. AFAC. Available at: <https://www.afac.com.au/initiative/fire-prediction-simulators> (accessed 15 June 2025).

134. Australian Institute of Marine Science (2025) Causes of crown-of-thorns starfish outbreaks. AIMS. Available at: <https://www.aims.gov.au/research-topics/environmental-issues/crown-thorns-starfish/causes-crown-thorns-starfish-outbreaks> (accessed 15 June 2025).
135. Chung G (2022) Partnering with CSIRO to protect our reef, with help from AI. Google. Available at: <https://blog.google/intl/en-au/company-news/outreach-initiatives/protecting-our-reef-with-csiro/> (accessed 15 June 2025).
136. Cranney K (2019) Magpie geese return with help from ethical AI and Indigenous Knowledge. CSIRO. Available at: <https://www.csiro.au/en/news/all/articles/2019/november/magpie-geese-return-ethical-ai-indigenous-knowledge> (accessed 15 June 2025).
137. MIT Technology Review (2023) Four ways AI is making the power grid faster and more resilient. MIT Technology Review. Available at: <https://www.technologyreview.com/2023/11/22/1083792/ai-power-grid-improvement/> (accessed 15 June 2025).
138. European Commission (2025) Corporate sustainability reporting. European Commission. Available at: <https://finance.ec.europa.eu/capital-markets-union-and-financial-markets/company-reporting-and-auditing/company-reporting/corporate-sustainability-reporting_en> (accessed 15 June 2025).
139. Alphinity Investment Management and CSIRO (2024) *The Intersection of Responsible AI and ESG: A Framework for Investors.* CSIRO Publishing, Melbourne.
140. Global Reporting Initiative (2025) GRI Standards. Global Reporting Initiative. Available at: <https://www.globalreporting.org/standards/> (accessed 15 June 2025).
141. IFRS Foundation (2025) SASB Standards. IFRS Foundation. Available at: <https://sasb.ifrs.org/> (accessed 15 June 2025).
142. Workforce Disclosure Initiative (2025) Workforce Disclosure Initiative. Thomson Reuters Foundation. Available at: <https://wdi.trust.org/> (accessed 15 June 2025).
143. Naughtin C, Hajkowicz S, Schleiger E, Bratanova A, Cameron A, Zamin T, Dutta A (2022) *Our Future World: Global Megatrends Impacting the Way We Live Over Coming Decades.* CSIRO, Brisbane.
144. Martin A (2024) Health insurance data breach affects nearly half of France's population, privacy regulator warns. *The Record*, 9 February.

Available at: <https://therecord.media/health-insurance-data-breach-affects-half-of-france-cnil> (accessed 16 June 2025).

145. Center for Strategic and International Studies (CSIS) (n.d.) Significant cyber incidents. *Strategic Technologies Program*. Available at: <https://www.csis.org/programs/strategic-technologies-program/significant-cyber-incidents> (accessed 16 June 2025).
146. Wu T, Yang S, Liu S, Nguyen D, Jang S, Abuadbba A (2025) ThreatModeling-LLM: automating threat modeling using large language models for banking system. *arXiv*. Available at: <https://arxiv.org/abs/2411.17058> (accessed 16 June 2025).
147. Cranko Z, Menon A, Nock R, Ong CS, Shi Z, Walder C (2019) Monge blunts Bayes: hardness results for adversarial training. In *Proceedings of the 36th International Conference on Machine Learning*, (Eds K Chaudhuri and R Salakhutdinov) pp. 9–15. Available at: <https://proceedings.mlr.press/v97/cranko19a.html> (accessed 29 July 2025).
148. Lees D (2023) Deep Fake Neighbour Wars: ITV's comedy shows how AI can transform popular culture. *The Conversation*, 27 January. Available at: <https://theconversation.com/deep-fake-neighbour-wars-itvs-comedy-shows-how-ai-can-transform-popular-culture-198569> (accessed 29 July 2025).
149. McGlynn C (2023) Cyberbullying girls with pornographic deepfakes is a form of misogyny. *The Conversation*, 9 October. Available at: <https://theconversation.com/cyberbullying-girls-with-pornographic-deepfakes-is-a-form-of-misogyny-217182> (accessed 16 June 2025).
150. Salam E (2023) US mother gets call from 'kidnapped daughter' – but it's really an AI scam. *The Guardian*, 14 June. Available at: <https://www.theguardian.com/us-news/2023/jun/14/ai-kidnapping-scam-senate-hearing-jennifer-destefano> (accessed 16 June 2025).
151. McGlynn C (2024) The dangers of voice cloning – and how to combat it. *The Conversation*, 13 June. Available at: <https://theconversation.com/the-dangers-of-voice-cloning-and-how-to-combat-it-239926> (accessed 16 June 2025).
152. Edwards B (2023) Microsoft's new AI can simulate anyone's voice with 3 seconds of audio. *Ars Technica*, 9 January. Available at: <https://arstechnica.com/information-technology/2023/01/microsofts-new-ai-can-simulate-anyones-voice-with-3-seconds-of-audio/> (accessed 16 June 2025).

153. Dezfouli A, Nock R, Dayan P (2020) Adversarial vulnerabilities of human decision-making. *Proceedings of the National Academy of Sciences of the United States of America* **117**, 292.
154. Callahan C (2024) How Amazon's generative AI tool for developers is saving 4,500 years of work, $260 million annually. *Digiday*, 22 August. Available at: <https://digiday.com/media/how-amazons-genai-tool-for-developers-is-saving-4500-years-of-work-260-million-annually/> (accessed 16 June 2025).
155. Roush T (2024) Microsoft claims Russia, China and others used OpenAI's tools for hacking. *Forbes Australia*, 14 February. Available at: <https://www.forbes.com.au/news/innovation/microsoft-russia-china-openai-tools-hacking/> (accessed 16 June 2025).
156. Zhong L, Wang Z (2024) Can LLM replace stack overflow? A study on robustness and reliability of large language model code generation. In *Proceedings of the Thirty-Eighth AAAI Conference on Artificial Intelligence, the Thirty-Sixth Conference on Innovative Applications of Artificial Intelligence and Fourteenth Symposium on Educational Advances in Artificial Intelligence (AAAI'24/IAAI'24/EAAI'24)*, vol. 38, Article 2437, pp. 21841–21849. AAAI Press, Washington, DC. <doi:10.1609/aaai.v38i19.30185>
157. Harmonic Security (2025) From payrolls to patents: the spectrum of data leaked into GenAI (Q4 2024 report). Harmonic Security. Availableat: <https://www.harmonic.security/resources/from-payrolls-to-patents-the-spectrum-of-data-leaked-into-genai> (accessed 16 June 2025).
158. Phil Muncaster (2024) Fifth of CISOs admit staff leaked data via GenAI. *Infosecurity Magazine*, 24 April 2024. Available at: <https://www.infosecurity-magazine.com/news/fifth-cisos-staff-leaked-data-genai/> (accessed 16 June 2025).
159. Weber K (2024) Companies fall behind on AI policies and training. *iTnews*, 14 May 2024. Available at: <https://www.itnews.com.au/news/companies-fall-behind-on-ai-policies-and-training-607893> (accessed 16 June 2025).

Index

www.ingramcontent.com/pod-product-compliance
Lightning Source LLC
LaVergne TN
LVHW101939220826
846093LV00006B/56

* 9 7 8 1 4 8 6 3 2 1 4 2 1 *